OPERATION DRAKE

Lieutenant Colonel John Blashford-Snell MBE, RE, was born in 1936. He is married and has two daughters. A serving soldier in the Royal Engineers, he is one of the world's most seasoned explorers. Since 1958 he has led or taken part in eighteen expeditions including major scientific exploration and underwater archaeological quests. In 1972 he led the British Trans-Americas expedition from Alaska to Cape Horn. His unique experience and qualities of leadership made him the ideal leader for Operation Drake.

Michael Cable, formerly the showbusiness correspondent of the *Daily Mail*, is a freelance journalist and author of three books on such diverse topics as the music industry and athletics.

DEDICATION

In memory of our good friends
Andrew Maara (Kenya) and
Richard Hopkins (Great Britain)
who died in road accidents in Kenya.
They epitomised the spirit of Operation Drake
with their selflessness, determination and
good humour.

IN THE WAKE OF DRAKE

Lieutenant Colonel John Blashford-Snell and
Michael Cable

A STAR BOOK

published by
the Paperback Division of
W. H. ALLEN & Co. Ltd

A Star Book
Published in 1980
by the Paperback Division of
W. H. Allen & Co. Ltd
A Howard and Wyndham Company
44 Hill Street, London W1X 8LB

Typeset by Computacomp (UK) Ltd,
Fort William, Scotland
and printed in Great Britain by
The Anchor Press Ltd
Tiptree, Essex
ISBN 0 352 30750 1

On behalf of all those who took part in Operation Drake I would like to thank those many friends, supporters and sponsors around the world who made the whole venture possible and the governments of Panama, Papua New Guinea, Indonesia and Kenya without whose understanding, co-operation and assistance it could never have been so successful.

J. B-S

Chapter 1

Policewoman Christine McHugh edged cautiously into the dimly-lit room and paused for a moment while her eyes adjusted to the gloom. At first she could see nothing particularly unusual – and that rather surprised her. She had been preparing herself for a shock. After all, nothing that had happened since she arrived at the spooky old priory had been predictable and she had found herself having to cope with situations the like of which she had never had to face before – not even on the beat in Gravesend! But that was fair enough. She had been well aware when she reported to St Augustine's Priory, in the Kent village of Bilsington, for a weekend with a difference that her courage and her initiative were going to be tested to the full.

She glanced again round the sparsely-furnished room and this time she noticed the sack – and the fact that it was moving in a decidedly sinister fashion. Her fingers nervously twisted the piece of string she had been handed a few moments before with the simple instruction to go and get the measure of whoever or whatever she found on the other side of the door in front of her.

She advanced towards the sack and was aware of a definite tingling up and down her spine as she fumbled with the rope that was tied round the top of it. Then, when she eventually loosened it and the sack fell open, she staggered back with a stifled gasp as the most enormous snake she had ever set eyes upon reared up and immediately started trying to wind itself around her neck.

His name, she later discovered, was Monty Python – but as

far as Christine was concerned, he was certainly no joke! After a considerable amount of grappling and grasping she managed to stretch her piece of string along the serpent's writhing length and to estimate that he measured at least 12 feet.

Establishing that fact was one of the trickier tasks faced by Christine and fifteen other young people who took part in the first of the testing Selection Weekends, organised in eight different centres throughout Britain – and also in Australia, Canada, New Zealand and the USA – with the aim of finally sorting out the lucky few who would be invited to take part in one of the greatest adventures of modern times – Operation Drake.

It had all started almost ten years before that bitter cold February weekend in 1978 when Christine and the rest of the short-listed candidates from the South East region found themselves competing not only against each other but also against the elements and their own powers of stamina, endurance, strength of character and calmness under pressure to win the once-in-a-lifetime prize of a place among the 216 Young Explorers to be selected from a total of over 58,000 enthusiastic applicants.

Sic Parvis Magna (Great Things Have Small Beginnings) was the motto of Sir Francis Drake and how aptly that notion could be applied to the unique round-the-world expedition which borrowed his name exactly 400 years after his epic circumnavigation in the Golden Hind between 1577–80.

The origins of Operation Drake can be traced back to the founding of the Scientific Exploration Society (SES) in 1969, although at that time it was no more than a vague idea that one or two of the founder members had in the backs of their minds. Gradually over the next three or four years if began to take shape under the somewhat unromantic heading of Project WW 895, and by the time the 150-ton brigantine *Eye of the Wind* sailed ceremonially out of Plymouth in October 1978, with Prince Charles at the helm to give the venture a right royal send-off, it had developed into one of the most ambitious, imaginative and wide-ranging expeditions ever mounted.

Although more than 2,000 people from twenty-six countries have been involved as Young Explorers, administrators, scientists, engineers, communications experts, fundraisers and sponsors and countless more have played unofficial, but often vitally helpful, roles behind the scenes. The four major land phases in Panama, Papua New Guinea, Sulawesi, and Kenya have featured important social, scientific, medical, archaeological and historical research projects as well as pioneering exploration and adventure. And the overall cost will eventually tot up to around £1 million in hard cash plus incalculable thousands of pounds worth of free and cut-price travel, goods and services.

Certainly nothing quite as grand as this was anticipated when the SES was first formed. In fact, the Society was really inspired by far more mundane considerations such as what to do with all the miscellaneous items of kit that had been accumulated during the twenty or so expeditions that I had launched as Adventure Training Officer at Sandhurst.

It was Operation Drake's Executive Chairman General Sir John Mogg – at that time Commandant at Sandhurst – who put forward the suggestion that what was really needed was an organisation with charitable status, through which future expeditions could be planned, equipped, administrated and, most important, financed. It made such good sense that six colleagues who had accompanied me on the successful Blue Nile Expedition met at my army home in Camberley and there, over a few good whiskies, the basic form, aims and ideals of the SES were agreed.

An international body of servicemen, scientists and explorers working closely with the British Museum and the Explorers Club of New York, it now has branches all over the world and a total membership of around 400.

Right from the very outset the long-term forward planning of expeditions was always high on the Society's agenda and the idea began to take shape of doing something rather special that would combine all the different facets of exploration and also exploit the full potential of the SES set-up.

From then on things seemed to develop naturally. The

possibility of actually linking together what had originally been planned as a series of separate expeditions in South America, Panama, the Galapagos Islands, Papua New Guinea and the Sudan began to emerge as an obvious development when it was pointed out that all the sites have a common factor in that they are approachable by sea. This meant that a ship could be used as a floating base thereby not only adding an extra dimension to the whole project but also serving the very useful purpose of enabling us to transport personnel, stores and equipment from one place to another in the simplest and most efficient manner.

Next came the most exciting and far-reaching expansion of the original plan – the decision to involve a large number of young people. Once again, the introduction of this major new element – which more than anything else helped to give Operation Drake its unique flavour – came about in such a way that is seemed a perfectly natural and logical progression.

We had been accompanied on the 1975 Zaire River Expedition by two teenage boys from my home island of Jersey. This resulted from an advertising executive friend of mine happening to mention one day that the Royal Trust Company wanted to celebrate its 75th anniversary but was not quite sure how to do it in a suitable fashion. It had been considering the idea of having 75 birthday cakes baked and delivered to poor people until it was pointed out that in a tax haven like Jersey millionaires are probably easier to find than poor people. John had then suggested that maybe it should think about raising £1,000 to send a young person on an expedition. Years of scraping together the funds to back our various ventures have taught us never to be slow in coming forward when faced with someone who apparently has money to spare. John's suggestion was seized upon with vast enthusiasm and almost before you could say 'Applications are invited . . .' hundreds of hopefuls had been screened, interviewed and tested and a young business student called Richard Le Boutillier duly selected. There was a bonus in that the police cadet he beat into second place was also so outstanding that it seemed a pity that he shouldn't be able to come along too and although there were problems involved in a policeman touting for sponsorship, I had a word with a friend of mine who was a senior officer in the Force and the money was forthcoming. It was well worthwhile from everybody's point of view because these young men proved an asset to the expedition whilst at the same time enjoyed a fantastic experience and from then on the inclusion of Young Explorers in our plans wherever possible became an accepted part of SES policy.

Not that we ever envisaged taking along quite so many as eventually joined us on Operation Drake until Prince Charles came up with the suggestion that maybe we should be thinking in terms not of just two or three youngsters but of two or three hundred. The idea of an expedition on such a large scale is frightening but it took very little consideration to realise that this not only made a lot of sense but also would broaden the outlook of the whole project and give it a much greater sense of purpose.

From our meetings with international youth organisations, schools and universities it had already become abundantly clear that many youngsters these days feel the need to be stretched and challenged more severely than they can ever normally hope to be in modern society. Outward Bound and similar adventure training schemes do their best to satisfy this need and to provide imaginative outlets for natural aggression and excess energy, but for a lot of young people they don't really go far enough. Again it was Prince Charles who pinpointed the underlying cause of their frustration and summed it up in a phrase, when he told the House of Lords that what modern youth was really seeking was 'some of the challenges of war in a peacetime situation'.

Inspiration was far from being the Prince's only contribution to the success of Operation Drake. Right from the very beginning he played a vital role as Patron and it is absolutely true to say that without his encouragement the project would never have got off the ground. He is by nature a man of action with a great spirit of adventure.

Now the only thing we needed was some firm indication that

the kind of money we required would be forthcoming. I say 'only' but even at that stage, before inflation and various unforeseen and expensive complications forced us to re-estimate drastically, we had priced the project at around £650,000. We reckoned we needed at least £50,000 up front.

This huge amount materialised in one lump sum in a spectacular manner. Mr Walter Annenberg, the former American ambassador in London, finds the whole idea of expeditions fascinating. He has done a great deal to help worthy causes and his generosity, foresight and imagination are well known. In view of this it was with a thrill of anticipation that I accepted his invitation to visit him at his home near Palm Springs the next time I was in the States to brief him on Operation Drake. As it happened, I was about to leave for America on a lecture tour and I wasted no time in fixing a meeting.

Along with my wife, Judith, and daughters, Emma and Victoria, I arrived at Palm Springs to be met at the airport by a chauffeur-driven limousine which whisked us to the estate, set like a green oasis in an arid desert where Mr Annenberg lives with Bob Hope and ex-President Gerald Ford as neighbours. From the moment we passed through the remote-controlled gates we were left breathless by the sheer scale and luxury of the place which is big enough to include its own private golf course.

We settled into a cottage in these grounds and I was just coming out of the shower when Emma rushed in to announce that Mr Annenberg had arrived to see if we were comfortable. I poured him a drink and without any real preamble he said, 'Tell me about this Drake thing – and how much you need to get it started.' I gave him a brief outline of what it was all about and announced as nonchalantly as possible that we required about £50,000 before we could really get underway. He asked me if I thought we could really pull off something quite so big and when I assured him that as long as we could get the finance there was no reason why we shouldn't be able to make it work, he paused for a moment and then said, 'If I give you

$100,000 – will that do you?'

I was flabbergasted. That was it. The moment I got back to England it was all systems go.

The nucleus of the Operation Drake team assembled in Room 5B, a dank, dark and rather dismal old storeroom in the bowels of Whitehall.

There was ex-Royal Engineer Captain Jim Masters, an old friend and veteran adventurer, who made such an impression on the local health authority during the Zaire River Expedition that they persuaded him to quit the army and stay on there for two years, totally re-organising the distribution and technical sides of the country's medical service. He came in as our head of logistics and personnel, a key role that involved arranging to move people and equipment around the world for nothing or next to nothing. Former company director Mervyn Price became Director of Administration in charge of keeping a close eye on our finances. This vital job was later taken over by George Thurstan, who resigned from the police force to join us. Ruth Mindel, who had already proved herself as a most capable PR lady, was brought in to combine the duties of Public Relations and Procurement Officer and another old friend, Val Roberts, valiantly signed on as Registrar in charge of the office. On the scientific side, Jerseyman Andrew Mitchell came to us from Bristol University to act as Scientific Co-ordinator and took responsibility for liaising between Freddie Rodgers' Scientific Sub-Committee and the general administration.

Our immediate task once we had got the green light to go-ahead was to start activating a worldwide network of local Operation Drake committees so that work could start on finding sponsors, selecting Young Explorers and, in areas where phases of the expedition would actually be taking place, preparing the way for our arrival. Messages flashed around the globe as we appointed a chairman here, a secretary there and so on. It all happened so quickly that Dr John Chapman-Smith, who was chosen to be our representative in New Zealand, was taken completely by surprise when he arrived back in Auckland after a brief trip abroad to be met by pressmen demanding to know more about Operation Drake. 'You tell *me*!' he exclaimed. 'I haven't the faintest idea.' Our letter explaining what it was all about and asking him to be our local chairman was still sitting in his 'in' tray at his office.

Meanwhile, we had a third major stroke of luck to follow our good fortune in securing Prince Charles' moral support and Walter Annenberg's financial backing when my former Commander from Sandhurst days, General Sir John Mogg, agreed to become our Executive Chairman. He was a little reluctant when I first approached him at his home in Oxfordshire, claiming that he was too old and had been looking forward to a quiet life in his retirement. But when I assured him – with an absolutely straight face – that the position would involve no more than two or three meetings a year he agreed that he could probably manage that.

Of course, things didn't work out quite as I implied and he ended up spending half his life tied up with Operation Drake. Nevertheless, I'm pretty certain he enjoyed every minute of it – even if it did seriously interfere with his fishing plans! And I was delighted to have secured the ideal chairman. As a former Deputy Supreme Allied Commander, Europe, his international standing and contacts are second to none and as a respected figure of the highest calibre he brought added prestige to the project. Furthermore, he had the rare ability of getting on terribly well with people and was really in tune with the young generation. Without him life would have been much more difficult.

By this stage things were beginning to run smoothly and the money was coming in steadily. A major boost came when I met John Whitney, the energetic Managing Director of Capital Radio with valuable sponsorships and publicity. Prince Charles was now confident that his association as Patron could be officially announced. He further agreed to launch Operation Drake formally on 13 December 1977 – the precise anniversary of Drake's sailing 400 years before. The actual start of the expedition was scheduled for October 1978 and the Prince promised to be at the helm when our ship *Eye of the Wind* left Plymouth.

For several days after the launching at the Athenaeum Hotel our HQ resembled a casualty station as we tried to cope with the avalanche of press, radio and television enquiries from all round the world – not to mention calls from youngsters wanting to sign on as Young Explorers (YE's). Capital Radio had over 10,000 enquiries from young people in the London area alone. Things went mad and when that initial excitement eventually died down a bit we were all hollow-eyed. One major consolation was that a considerable number of the inquiries we had to deal with were from people wanting to contribute to our funds and when we eventually counted up the money, postal orders and cheques that cascaded in from ordinary members of the public they totted up to a £58,000 bonus.

All over the world, hundreds of good folk gave time and effort to help us. The Managing Director of ICI Australia, Denis Cordner, had already started to form committees throughout his country, whilst in the USA Sir Gordon Whote of the Hanson Trust became Chairman and with Gordon Booth (later also to be knighted) as Secretary formed a powerful team. As ever, Otto Roethenmund and Vince Martinelli were in the forefront of the organisation and we were greatly helped by having Mr William Seawell, President of Pan Am in the team. Pan Am's generosity made all the difference. They kindly flew many of our staff and all the US Young Explorers and they also allowed Barbara Martinelli to join our medical research team.

In Nepal it was businessman Jim Edwards, in Papua New Guinea an old schoolfriend, Nigel Porteous, gave a hand, and in West Germany Captain Dick Festorazzi of the Royal Engineers set up a committee in Willich. In America Dr Dan Osman, a tried and trusted ally, agreed to recruit medical staff for us and help with fundraising, whilst New York businessman John Linehan kindly gave us full use of his office and great assistance in such matters as the selling of the film of the expedition to American television. Squadron Leader David Checketts, who was then Prince Charles' Private Secretary, became Chairman of our Publicity Committee. Our old friends Gestetners gave us an office in London and New York and facilities throughout the world, as well as a new inflatable diving tester. BP generously provided the fuel for the ship, Esso gave us our financial control organisation and auditors. BAT offered help in all the major land phases and, through the good offices of Robin Leonard, provided major back-up in Indonesia. The Avon Rubber Company gave a supply of their tough inflatables. Airlines, including British Airways, Cathay Pacific, Air Nuigini, Quantas and British Caledonian, offered valuable assistance. In no time we had over 300 sponsoring companies and Ruth Mindel's department was taxed to its limit in looking after this army of supporters – so much so that we asked Neilson McCarthy, the well-known Public Relations Company, to take over our publicity.

Without a doubt our greatest single supporter was the Armed Forces, whose members took part in Operation Drake under the Adventurous Training scheme. Without the support of the Royal Signals I doubt if our communications would have coped with the vast distances and variety of terrain that we were to experience.

By early 1978 we were ready to start making the final selection of YE's of whom there were to be twenty-four on each of the nine phases of the expedition. Of the basic total of 216 YE's – the addition of various 'special cases' along the way eventually swelled this figure to nearer 300 – about 100 came from Britain. And for these precious few places we received altogether more than 50,000 applications. After careful consideration about 8,000 were chosen for personal interviews as a result of which some 1,500 went onto the short list of candidates who were invited to take part in one of the Selection Weekends – between twenty-four and thirty-six hours of concentrated tests carried out in a carefully-contrived context of hardship and discomfort.

The first one in Kent was probably one of the toughest if only because the conditions were so utterly miserable. The weather couldn't make up its mind whether to rain, sleet or snow and it was bitterly cold as the sixteen candidates assembled at 11.00 am on the Saturday for a preliminary briefing. After a kit inspection, to make sure they had brought along everything they had been told to bring, they were marched into the woods, issued with tents and given thirty minutes to set up camp. They were then divided into two teams of eight and taken to a nearby pond for the first test which had been designed with the sadistic intention of trying to ensure that everybody experienced the maximum discomfort at the earliest possible stage of the proceedings. They were told to imagine that they were being pursued through the jungle by hostile natives, that one of their number was badly wounded and that they had just one hour to get across the crocodile-infested river represented by this murky Kentish pond. They were given a highly unstable single-seat canoe and some oil drums and rope with which to construct a makeshift raft for their casualty. The idea was that they should all fall in and get soaked to the skin and most of them duly obliged. I must confess that it sent shivers up my spine to watch them splashing about in the icy water. The totally incongruous background sounds of tropical jungle wildlife, which we had recorded and played at ear-splitting volume through speakers hidden in surrounding trees and bushes, could have done nothing to warm them up!

It was after this ordeal that they were led off for their confrontation with Monty. The aim here was obviously to make sure that they weren't too nervous of creepy crawlies – a terrible disadvantage for anybody contemplating a spell in the jungle. We had originally planned to face them with an 18 foot radio-controlled crocodile which Ruth Mindel had somehow unearthed but there turned out to be problems about transporting this mechanical beast on the roofrack of her Mini! An inflatable dinosaur was briefly considered before Andrew Mitchell discovered a firm which specialised in providing unusual creatures for every occasion – mostly for films and television. After sifting through their catalogue of exotic and versatile beasts we settled for Monty Python.

He frightened most of the Directing Staff half to death the first time we saw him – and *we* knew what to expect! So I can well understand why there were so many stifled screams from the YE's when he suddenly came slithering out of his sack or from under the pile of carpets where we occasionally hid him. He was actually quite harmless – at least, he was most of the time. There was just one dramatic moment when he did get out of control and caused some concern.

It happened at the Wembley Conference Centre during a special Capital Radio presentation aimed at showing what Operation Drake was all about. Peter Shea – one of the four successful YE candidates from that first Selection Weekend – had agreed to go on stage to put Monty through his paces. By that time we had slightly modified the test so that instead of measuring him the YE's had to weigh him using only a small pair of bathroom scales. This had led to some hilarious attempts to roll him into a ball and make him stay that way while the scales were read, and one youngster actually tried to stand him on his tail and then let go getting a reading before he fell off. Monty suffered all these indignities without putting even the slightest squeeze on anybody until poor Peter went up to demonstrate the correct way of doing it – namely, by weighing yourself first alone and then again with Monty round your shoulders and subtracting one result from the other.

Maybe it was the heat of the footlights that made Monty particularly lively that day by reminding him of his natural habitat – I don't know. But whatever it was it soon became clear that Peter was in trouble because his voice, as he gave a running commentary on what he was doing, began to get more and more strained as Monty wound himself ever tighter around his neck until all that came out was a croak! Fortunately the handlers were close by and they rushed on stage and hurriedly unwound the unruly serpent whereupon Peter sank red-faced and somewhat breathless to the floor.

Monty wasn't the only 'frightener' we used to test the candidates' nerve. At one Selection Weekend held at the Jersey Zoo candidates were asked to roll up their sleeve, hold out their hand and close their eyes, whereupon a vast hairy-legged tarantula spider would be sent crawling up their arm. It was, of course, a non-poisonous variety but needless to say we didn't go out of our way to disclose that fact!

By the evening of the first day of the Selection Weekend the candidates would be very cold, very wet, very tired and ravenously hungry, but dreams of being handed a nice supper on a plate were dashed when they were handed a few brace of pigeons – unplucked, naturally – and left to their own devices. If they were lucky they would get a few mouthfuls of half-cooked, half-burnt flesh before collapsing into their sleeping bags.

An hour later they would be rudely awakened and despatched on a cross-country march to a map reference where they were warned to expect some kind of emergency only to find when they eventually arrived there that it was a false alarm. After trudging all the way back and crawling, even more shattered, into their sleeping bags, they would be dragged out yet again at about 3.00 am and sent marching off to another map reference. This time they *would* find some kind of staged emergency waiting for them and would have to gather their scattered wits sufficiently to deal with it in a sensible fashion.

Then it would be back to camp for perhaps a few minutes of shut-eye before we woke them at 6.00 am with breakfast – a fresh mackerel that would land with a thud on the ground outside their tents. An easy thing to deal with in a modern kitchen, but not quite so straightforward when all you have is a knife and fork, a small mess tin and a box of matches. In most cases the results of the candidates' culinary endeavours tended to be less than mouth-watering.

So for most people it was an unappetising start to the second day's programme of tests which were aimed at sounding out mental, rather than physical, capabilities and, in particular, the candidates' powers of concentration and original thought while in a state of considerable exhaustion. The prospect of having to get up in front of an audience and speak for three minutes on any chosen subject proved just as demanding for many people as any of the hardships that had gone before.

At the end of the weekend we would have a pretty good idea of exactly what each candidate was made of. We didn't necessarily choose those with the top marks because that way we might have ended up with a lot of the same types, whereas what we were really after was a well-balanced, mixed bunch. We wanted to be able to put together teams that might include, for example, someone highly intelligent and about to take his medical finals on the one hand and, on the other, a kid from a broken home who had been in trouble with the police and maybe even on the verge of Borstal. That meant looking for people whose talents and personalities would complement each other.

There was really only one qualification that was essential – a spirit of adventure.

Chapter 2

On Sunday 22 October 1978 the *Eye of the Wind* duly left Plymouth harbour and headed out to sea right on schedule to get Operation Drake off to what had all the appearances of a very spectacular start. It was a magnificent sight, such as might bring a lump to any old sea dog's throat, as the gleaming white ship glided gracefully past the Hoe under full sail with Prince Charles at the helm. Guns were fired, the crowds cheered and the armada of small boats that escorted her whoop-whooped their farewells. Stirring stuff indeed – and a great pity, really, that it was all just a grand illusion!

The ship didn't actually sail out that day – she was towed out by a naval tug using a very long line so that it appeared that she was making her own way. And once Prince Charles had been taken off by launch and then whisked away by helicopter, and once the film and TV cameras had stopped rolling and the crowds had all gone home, we towed her back in again. To those who noticed we gave the explanation that she had to undergo sea trials before she could actually start the voyage. The sad truth was that the ship was far from ready to sail. Her new engine, largely paid for by Prince Charles, was not ready to run and there was still some refitting work to be done and equipment to be loaded. It had been a frantic race against time to get her into any fit state even for that brief ceremonial trip along Plymouth Sound and back again. There was even a story going the rounds that her hull had only been painted white on the side that faced the cameras and the crowds and that on the other side she was still the dirty black she had been when we took delivery of her. That wasn't actually the case – although it was uncomfortably near the truth. The deck had certainly not been painted – that chore wasn't completed until they reached Tenerife.

I have often thought of Operation Drake as being like a two-year hurdle race during which we were constantly having to clear a seemingly endless succession of barriers and on that analogy there is no doubt that on every lap the most insurmountable obstacles we had to negotiate were those connected in one way or another with the ship.

Just trying to find a suitable vessel in the first place was difficult enough. Our first choice had to be ruled out after my aide Chris Sainsbury – destined to become Operation Drake's invaluable roving Jack-of-all-trades, chief photographer and resident 'character' – went for a trial voyage aboard her and reported back that she leaked like a sieve, was hard to sail and shouldn't be touched with a bargepole.

Our next alternative was the *Captain Scott* – a fantastic modern sailing ship only built in 1971 and owned by a Trust. They agreed to charter it to us for just one pound – the catch being that we had to foot a maintenance bill that we reckoned would add about £100,000 to that. Even so, it was still a good deal for a ship like that and so we decided to go ahead. But at the last minute the Trust sold it instead to the Sultan of Oman, good business from their point of view, no doubt – but for us it was a disaster. Part of the deal had always been that the Trust reserved the right to sell the ship if they got a good offer before the date when it was due to be handed over to us, but we never dreamed that this would happen. It was against all the odds. Now, when all our other plans were advanced beyond the point of no return, we suddenly found ourselves without a ship. It was our first major crisis.

A frantic search for a replacement began and just when it began to seem that we would never be able to find what we needed, Fate stepped in. Out of the blue I received a picture postcard of a brigantine called *Eye of the Wind* from an American businessman, Allen O'Brien. On the back was a note informing me that he was now co-owner of the vessel, that he had read in a Scientific Exploration Society Newsletter about Operation Drake and that if we should ever have need of a Tall Ship on that or any other expedition we should bear the *Eye of the Wind* in mind. Such a timely offer was almost too good to be true and I wasted not a moment in contacting him with the result that the ship was chartered to us for a very reasonable £1,000 per month. But if we thought our troubles were over once we had acquired the 132-foot, 150-ton square rigger we were very quickly disillusioned. From the moment she came into our hands she posed a never-ending succession of nightmare problems that threatened not only our sanity but our solvency as well.

The *Eye of the Wind* admitted to a functional and rather uninspiring history. She was built in 1911 at Braca, near Bremen, specifically to ply the round trip from Germany to Argentina to Cornwall and back to Germany carrying salt from Germany, hides from Argentina and china clay from Cornwall. What happened to her during the First World War is not known but she re-appeared in 1923 when she was bought by a Swedish concern and for the next forty-five years she spent her winters trading cargo in the Baltic and the North Sea and her summers drifting for herring off Iceland. In the late sixties she was badly damaged by a fire which broke out while she was ice-bound in the Baltic and after being towed back to Gothenburg she was left there to rot until the present owners – who had been searching for some time for just such a ship – found the burnt-out hulk and decided to buy her.

The syndicate of six – tragically reduced to five when Allen O'Brien was killed in an air crash in India in 1980 – spent six months patching her up sufficiently to get her from Sweden to England and after a one-year interlude, while they raised the money to refit her properly, they spent eighteen months painstakingly restoring her using their own hands and some volunteer labour. That in itself was quite an achievement which was further remarkable for some fascinating examples of improvisation.

The floor of the lower saloon came from a dance hall that was being converted for bingo, the bench seats were actually pews from an old church and the magnificent oak panelling was salvaged from a bank that was being pulled down. The superb teak in the deckhouse was rescued from an Australian minesweeper that was being broken up, while the compass was taken from an old trawler that had also ended its days in the breaker's yard. The pinrail was fashioned from old sleepers that once supported the Tenterden railway in Kent and the spars were made by the owners themselves using adze and spokeshave in the old-fashioned way.

Despite all this admirable work by the syndicate – led by Londoner 'Tiger' Timbs who stayed on board as mate throughout Operation Drake – the *Eye of the Wind* was a long way from being up to the standard we required when she was finally delivered to us in Plymouth in August 1978. In fact, when we went into it in detail, it became alarmingly clear that we were going to have to start virtually from scratch and rebuild her. For one thing she needed a new engine and for another we were left with no alternative but to carry out fairly drastic alterations below decks in order to provide adequate accommodation. And those were just the major items! There was a frighteningly long list of less fundamental, but equally necessary, repairs and refinements. What's more, just when we thought we had everything covered somebody always seemed to unearth another unforeseen complication. The loo pipes were a typical example. Because Royalty was going to be on board – albeit only for a matter of minutes – the Department of Trade and Industry inspectors quite understandably made sure that the safety regulations were followed to the letter and among other things this meant ripping out all the plastic plumbing at the last moment and replacing it with copper fireproof piping.

Once it was realised that in addition to the astronomical extra cost of carrying out all the refitting there was apparently no way in the world that it was going to be completed in time, gloom and despondency – not to mention downright panic – set in. We were saved in the end by Patrick Collis, the man we had chosen to captain the ship, who persuaded his old Naval chums to help get the ship ready. We also got fantastic help from HMS Drake. Quite apart from speeding things up this move also saved us a considerable amount of money. The *Eye of the Wind* was still late making her final departure – but not disastrously so. And we didn't, after all, have to cancel the expedition's Royal launching day which would have been a terrible disappointment for everybody concerned and might well have been seen as an ill omen.

Against all the odds the big day went off without any major hitch – although there were a couple of anxious moments. The first involved the flying of the Royal Standard on the *Eye of the Wind* while Prince Charles was on board. The Royal Navy had specifically asked that we should not fly it until the ship had left the dockyard, since this would oblige every ship there to parade bands and guards and, as it was a Sunday, this would mean many sailors missing their leave. We agreed and the flag was left furled and stowed under the mast.

As the *Eye of the Wind* left the harbour I was following in a launch along with General Sir John Mogg, the Lord Lieutenant of Devonshire, Field Marshal Sir Richard Hull and the Admiral and I was just thinking to myself how smoothly everything was going when I noticed with alarm that the Admiral's knuckles were whitening as he clutched the brass rail in front of him while simultaneously a flush was spreading upwards from his collar – classic and immediately identifiable symptoms of suppressed fury. I followed his stony gaze and experienced an awful sinking feeling as I recognised the Royal Standard fluttering in the breeze. I learned later that the Prince had hardly got on board before he was asking Captain Collis, 'Where's my Standard then?' When Patrick explained why it was not being flown, HRH paused for a moment and then grinned mischievously and said, 'Well, let's break it and see if they've got any bands and guards!'

The second rather embarrassing incident would have appealed, I'm quite sure, to the Prince's sense of humour since it was like something straight out of a Goon Show script. A vital point of Royal etiquette is the provision of special private toilet facilities wherever an official visit is being made and we had not overlooked this. A particularly handsome prefabricated edifice had been erected by the Department of the Environment near the helicopter landing pad and sure enough, just before boarding the aircraft to leave, Prince Charles decided to avail himself of this convenience. As he advanced towards it an officer jumped smartly forward to open the door only to find it apparently jammed. Increasingly harder pulls failed to shift it whereupon a burly Petty Officer came forward and virtually braced one foot against the door jamb but still failed to budge it an inch. Quite obviously it was locked. The Prince observed all this with some amusement before murmuring, 'I do hope the key turns up' and departing. It emerged later that the guard who had been thoughtfully posted had been under the impression that if the building was not required when the Prince first landed it would not be required at all and had consequently gone off early with the key in his pocket.

For that individual it was, I suspect, a day he would prefer to forget. But it was a memorable occasion for most of the rest of us – particularly the YE's aboard the *Eye of the Wind* who were delighted to find Prince Charles in a charming, informal and relaxed mood. For a short time he took the wheel of the ship and at one point turned with a grin on his face and remarked casually, 'All I need now is a parrot'. He chatted to them for about two hours and even posed for pictures with young Christine McHugh who insisted on getting a profile shot as well. 'I wouldn't recommend it,' murmured HRH – but perseverance is one quality that all good YE's should have and WPC McHugh got her snap.

In the launch on the way back to shore the Prince was bubbling with enthusiasm about the whole venture and it seemed clear that his imagination had been fired by his contact with the ship and the YE's. For everyone in our HQ it was a relief to see something go right after so many weeks of crisis. But if we thought it was all going to be plain sailing from there on the *Eye of the Wind* soon produced the kind of nasty shocks that put paid to that fond notion.

Drake himself recorded in his log that on his first day out back in 1577 he was driven back to shelter by a tempest in which he lost his mast and, although the *Eye of the Wind* didn't suffer quite as badly as that, she certainly ran into a violent storm and spent her first night at sea tied to a buoy, being battered and buffeted in a way that had just about everyone on board reeling and retching. Few of the YE's escaped this early introduction to the miseries of really severe sea-sickness and by the time they arrived in Jersey most of them were still looking a trifle green about the gills and rather shaken. But they soon recovered to enjoy a great welcome that had been prepared by the local Operation Drake committee under the energetic chairmanship of my old schoolfriend Adrian Troy. My enjoyment of this was spoiled by the news with which the skipper greeted me, which was that after only a few hours at sea it was obvious that the ship was not 100 per cent right. The gear box had developed an extremely expensive-sounding knocking noise and there was something drastically wrong with the anchor chain which was causing it to slip alarmingly whenever it was raised, making this a slow and potentially dangerous operation. We decided that there was nothing we could do at that stage because, apart from anything else, we couldn't afford to get any further behind schedule. The winds looked to be set fair for the Atlantic crossing so with any luck the engine wouldn't be needed that much and once in Panama there would be the breathing space in which to sort things out.

In fact, the gear box packed up completely before the ship reached Colon with the result that the only way to shift gear was to get the revs exactly right and then jump heavily on the gear stick in the engine room. There was nothing for it but to fly the YE's on to our base camp at Caledonia Bay while the *Eye of the Wind* waited for a new gear box to be flown out from England – at a cost in air freight charges of £3,000! And this was at a time when we were so tight for money that when another £2,000 YE sponsorship came in, the celebrations in 5B were reminiscent of the relief of Mafeking!

The basic problem with the ship, it now emerged, was one of new wine and old bottles. In an old vessel like that there is no such thing as a standard part and when the new engine was put in a considerable amount of improvisation was necessary in order to marry it up to the existing prop shaft and gear box. Unfortunately the alignment was not quite right and so we successively broke one thing after another. After the gearbox it was the turn of the prop shaft to start giving trouble and that finally gave up the ghost in the Pacific. The wrong alignment meant that it was spinning through an oval rather than a circular revolution and inevitably this gradually got worse and worse until it eventually damaged the shaft housing. What's more, the vibration had been shaking the propeller itself loose and we only avoided losing it altogether by an amazing stroke of luck and a fraction of an inch. In Fiji divers went down for a routine, below-the-waterline inspection and discovered that the propeller was only held on by one bolt – and even that was down to the last couple of threads! Another few hours motoring at the most and it would have dropped off. Even then there was still a danger that we might have lost it because to adjust it and bolt it on again underwater was an incredibly tricky operation. And if it had gone we would have been in real trouble because that propeller is unique to the *Eye of the Wind* and a replacement would have been impossible to come by.

Back in Panama the problem of the anchor chain had required yet another drastic solution. Here we had another lucky break in the timely arrival on the scene of Army Air Corps helicopter pilot Major Frank Esson, who joined the expedition as senior watch leader aboard the *Eye of the Wind*. Frank is one of those marvellous mechanically-minded people who also had a great advantage over the rest of the Operation Drake directing staff in that he knew a lot about ships. It didn't take him long to confirm that we had the wrong kind of chain which was slipping because it did not fit the gypsy around which it had to be wound in. There was no alternative, once more, but to get a replacement and that, we were well aware, was easier said than done. It was also likely to break the bank yet again. The story of how Frank Esson managed to find a new chain and get it to the ship for virtually nothing has already passed into expedition folklore.

There are only a handful of chain yards in the entire world where he could expect to find what he was looking for and

Frank set out quite prepared to try all of them if necessary, starting with the nearest one in New Orleans, moving on up the coast to New York, then across to San Francisco and, in the last resort, right back to England. That prospect was quelling enough in itself, but his travel arrangements were further complicated by one enormous handicap – he had to take the gypsy with him! There was no way round this – it was the only way he could ensure that he got a chain that fitted. And just to make things even more difficult, he had to get the thing on the plane with him as personal baggage because we hadn't got the time to air freight it from one place to the next. Since one man could scarcely lift the 190 lb lump of ironmongery off the ground this was clearly going to stretch Frank's ability to charm airline staff to the very limit.

Anyway, there was no time to be lost because Easter weekend was approaching and the *Eye of the Wind* was due to sail on the Tuesday, so no sooner had the decision to search out a replacement been arrived at than Frank was on his way. It was all so sudden that there wasn't even time to go to the bank and get some money for him, so we had a quick whip-round which yielded no more than a few dollars. Frank's luggage consisted of a toothbrush – and this damned great gypsy which he had crammed into his hold-all.

The first stage of his journey – from Panama to Miami – presented no great problem. We had discovered by that time that if you wore uniform and looked reasonably respectable you could bluff your way past almost any official in Panama, so he and George Thurstan simply marched into the airport and made a big fuss about how this vital bit of equipment belonging to Prince Charles must go on the plane without fail. Eyebrows shot up when Frank's innocent looking hold-all sent the indicator on the baggage scales right off the clock but no objection was raised.

It was a different matter when he got to Miami and tried to arrange a connecting flight to New Orleans. After twenty hours and ten airlines he eventually conned someone into accepting him and his luggage. It must have been very funny to watch. The airline desks were all situated in the same concourse and after the first couple of refusals Frank developed a system whereby he would push the hold-all along the marble floor until he got to a pillar near the desk he was planning to approach next. He would then pause for a few moments to catch his breath and summon his strength and then lift the bag and waddle the last few feet as nonchalantly as possible. Nobody was fooled and, in fact, Frank was aware that his progress up the concourse was being watched with great amusement by those who had already turned him down.

He finally arrived in New Orleans late at night and because he had so little money with him he slept on a bench in the bus terminal, cuddling his precious gypsy for safety – though heaven knows who would have wanted to steal a thing like that. As soon as the yard opened at 6.00 am the next day Frank fell in through the gate and after rummaging around for an hour or two unearthed a chain that was a perfect fit. One final effort was then required to get the several tons of chain aboard the Panama Canal Company ship *Cristobal* before she sailed that afternoon. He managed to do that and so it was that the *Eye of the Wind* began her voyage across the Pacific right on time complete with a new chain. As an added bonus the Panama Canal Company decided that since they were about to give away the canal itself it really wouldn't make that much difference to them if they were to donate a chain to Operation Drake and so we saved ourselves several thousand pounds into the bargain. Frank spent a couple more uncomfortable nights on bus and airport benches, having used just about his last dime to telephone us with a request to organise his ticket back to Panama. He informed us when he returned that he had been forced to live for three days on nothing but beer but by the look on his face he had not found that too much of a hardship!

In Panama a new skipper took over *Eye of the Wind*. Mike Kichenside flew from Australia with his daughter Miranda, and they quickly became firm favourites with everyone. So a land phase ended and one of the sea phases began.

At this point I returned to Britain to catch up with my Army job and, as ever, to raise more money for the bottomless pit that Operation Drake seemed to have become.

While the series of crises during the very early part of the expedition caused understandable concern in the anyway rather gloomy depths of our London HQ, it did little to spoil the enjoyment of the YE's aboard the *Eye of the Wind*. After the storm-tossed and stomach-wrenching start to their voyage, the first day out of Jersey was idyllic with the ship drifting along at a pleasant two knots in the warm November sunshine. Chris Sainsbury noted in his log, 'The weeks of being messed about in Plymouth had taken a heavy toll of the morale of the whole ship's company but when we sat down to our lunch today – the first meal that had not come up again over the lee rail! – I could sense the most remarkable atmosphere of comradeship which until this point had been severely lacking. None of the YE's quite knows how life at sea will affect him or her but there is a great feeling of expectation.'

They didn't have to wait long to find out what it was all about because on the second day out from St Helier the wind got up again and after the sheer exhilaration of slicing through the water at eight knots under full sail there followed the thrill of going aloft in the moonlight to change sail, swaying about in the rigging while the captain and the bosun below bawled orders from the poop and foredecks. Even those with no head for heights admitted that once they got up there their nervousness seemed to vanish as the excitement of such a rare experience took over.

The strong winds developed into gales and sea-sickness returned for a few days before people eventually found their sea legs. One felt a special sympathy for the two Nepalese YE's Dilliram and Mohan Limbu who were particularly badly affected and yet, because of their very limited English, had to suffer in silence. Once the miseries of queasiness had been conquered it was a matter of getting used to the minor inconveniences of life aboard a sailing ship such as coming off watch at 4.00 am to find a soaking bunk waiting for you – or watching a lovingly prepared dinner disappear overboard, as happened once when the *Eye of the Wind* lurched violently into the trough between two waves so that water swirled across the deck and swept away two large baskets full of chicken legs and laboriously-peeled potatoes.

Such small irritations were, almost literally, like water off a duck's back to a bunch of adventurous youngsters and spirits were high enough to indulge to the full in the practical jokes, parties and 'variety' shows that traditionally relieve the monotony of a long voyage. The entertainment high point of the trip was the show laid on halfway across the Atlantic as a substitute for a crossing-the-line celebration. The various watches worked for days on their songs and sketches – one lot actually called up a passing naval ship on the radio in order to obtain the full unexpurgated lyrics of 'The Good Ship Venus' – and the evening, in the planning of which senior Capital Radio producer David Briggs naturally played an important role, was a riotous success.

Two of the lady YE's were victims of wildly successful practical jokes. Devonshire lass Mary Newman was one night in charge of hauling in marine biologist Patricia Holdway's plankton nets when, in the dim deck lighting, she noticed a real denizen of the deep trapped in the mesh. Little did the poor girl realise that it was a plastic octopus maliciously placed there when she wasn't looking by Chris Sainsbury and 'Tiger' Timbs. After rushing off to consult Patricia, Mary returned wearing an enormous pair of rubber gloves and clasping a pair of tongs with which she deftly removed the monster and placed it in a bucket for closer inspection. The rest of the watch, all of whom were party to the joke, could scarcely contain their glee as Mary was heard to announce triumphantly to Patricia that it couldn't possibly be an octopus as it had more than six legs. At this point Sainsbury stepped forward and with sadistic delight pointed out the 'Made in Japan' label whereupon he discovered that young Mary packed a mean right uppercut.

Much more elaborate and long drawn-out was the trick played on West Country policewoman Kirsty Macdonald-Henderson. It all started when she remarked that a cloud formation on the horizon looked very much like land and watchkeeper Peter 'Spider' Anderson from Australia picked up on this and told her very confidentially that it was the coast of Africa and that the ship had sailed in a complete circle. He went on to spin an incredible yarn about how she mustn't tell the others but this was all to do with the fact that they had strayed into the Cape Verde Oblong – an area of the ocean rather like the Bermuda Triangle where strange and inexplicable things happened. From then on Spider and one or two others organised all sorts of strange manifestations. The wheel would be mysteriously jammed, the engines and the radio would behave in a weird fashion and the electrics would go on and off for no apparent reason. The climax to the joke came one night when things were arranged so that Kirsty was up on deck alone. Suddenly she heard a strange noise and there, crawling on board over the gunwhale, was a sea monster that would have done credit to a Hammer horror movie. As it lurched towards her all dripping and slimy with the moonlight glinting on its hideous features Kirsty turned and fled. Only then was the awful beast revealed as Spider dressed up in a frogman suit and festooned with seaweed and other convincing accessories.

The most was made of a one-night stopover in Barbados with a visit to the notorious Harry's Nitery and, although there are those who swear that the place has never been quite the same since Harry himself passed on, most of the YE's nevertheless staggered out wide-eyed after witnessing sights beside which everything else they subsequently experienced on the expedition – including even the most bizarre mysteries of the uncharted jungle – paled into everyday insignificance!

The next stop was St Vincent where the crater of the volcano La Souffrire was to be studied. It was on Christmas Eve that the YE's lugged an Avon inflatable dinghy up the side of the volcano to the rim of the crater and the sight that greeted them made all the sweat worthwhile. So deep was the crater that their view of the lake and the central island at the bottom was partly obscured by clouds below where they stood. They had only a few minutes to admire the scene before applying themselves to the task of getting the Avon down to the water's edge which, owing to the almost sheer slope down, was achieved in less than totally elegant style with much slipping and sliding. The main scientific task was to make the first proper thermometric study of the crater for more than five years and the results they got gave alarming indications that a new eruption was imminent. Sure enough, three months later, it happened. So Operation Drake had the satisfaction of knowing that something truly practical had been achieved in that we were able to give an early warning of a potential disaster.

There wasn't time for any real Christmas celebrations as the *Eye of the Wind* set off on Christmas Day in the morning on the last lap of her journey. New Year's Eve also passed quietly with just a few sips of champagne and a rendering of 'Auld Lang Syne' – but clearly not every crew in the area was content to welcome in 1979 so soberly.

At about 1.00 am a ship was sighted on the horizon and as she approached nearer and nearer it became clear that she was not only on a collision course but showed no sign of being about to alter her bearing. The *Eye of the Wind* illuminated her sails with a powerful searchlight so that she must be visible to the other vessel but still nothing happened. As a last resort Radio Officer Corporal Roger Secker was ordered to call her up and remind her in no uncertain terms that, as a sailing ship, the *Eye of the Wind* had right of way and would she kindly take evasive action. At the very last moment she did alter course so as to pass just under the stern of the *Eye of the Wind* but then immediately turned round and headed back straight towards her. Slight unease now gripped Chris Sainsbury on the bridge and he altered course – only to find the other ship following his every move. Just when he was beginning to think that maybe he was dealing with one of the pirate ships that are more common on the oceans of the world today than they ever were in Drake's time, she called up on the radio, wished a 'Happy New Year' and disappeared.

The next day there was excitement of a different kind as land was sighted and the eager YE's, straining their eyes to the horizon, got their first glimpse of the jungle.

Chapter 3

Every minute of every day another 50 acres of jungle disappears for ever as modern logging machinery relentlessly mows down the mighty trees at the rate of 500 tons per hour. It is estimated on this basis that the world's lowland rain forests will have vanished completely within the next twenty years. In a world that is belatedly growing very anxious about its dwindling natural resources these statistics are perhaps the most alarming of all – far more frightening even than the speed at which the oil wells are drying up or the scale upon which various species of animal are being wiped out. After all, man will undoubtedly find a substitute for oil and there is hope, theoretically, for any creature as long as just one pair remains from which to breed. But for some reason a tropical rain forest, once cleared, never grows up again in quite the same way.

The explanation for this is apparently to be found in the fact that the flamboyant luxuriance of the jungle does not, as one might suppose, indicate exceptionally rich and fertile soil. In the warm, moist conditions the whole cycle of plant life is speeded up and the turnover of nutrients is much more rapid than in a temperate environment. When a tree dies decomposition takes place very quickly and the goodness is absorbed instantly by new and existing plants hungry for the food to fuel their hothouse development. The result of this is that the natural goodness necessary to plant life is retained not in the soil but in the plants themselves – in particular the trees. Remove the trees and you remove the nutrients as well leaving behind land which is good for nothing but scrub farming –

largely infertile and open to the ravages of erosion.

The wholesale destruction of the great forests around the world threatens society at every level. In the grand global scheme of things one has to remember that we rely on trees for our supplies of oxygen, while it is also accepted that the tropical rain forests play an important role in mitigating climatic effects and regulating the flow and quality of water – one fifth of the world's fresh water is locked in the Amazon basin alone. To remove such a large piece from the universal jigsaw puzzle seems like just another example of man's irresponsibility.

If people in the industrialised world find such considerations a little remote, a list of some of the everyday benefits provided by the jungle which have come to be taken for granted should help to bring home to them that the forest lands should not any longer be regarded casually as dispensable wilderness. Top of that list would be drugs such as the anaesthetics based on curare – first refined from South American jungle plants by the Amerindians – or the raiuwolfia from South East Asia's forests which plays such a vital part in modern heart surgery. Life would be less comfortable as well as less colourful without the steaming jungles.

On a local level, native people who have for centuries lived in perfect harmony with the forest are suffering directly from their destruction. When the protective vegetation is cleared from the hillsides droughts, flooding and landslides occur. And disease sometimes follows as organisms previously confined to the tops of the trees begin to infect humans on the ground.

It was with all this very much in mind that Operation Drake decided to base three of its four major land phases in jungle regions – first Panama, then Papua New Guinea and thirdly Sulawesi. It made sense to give our scientists the chance to do research in areas where the thrill of breaking completely new ground could be combined with the incentive of knowing that their findings could have invaluable practical significance.

For botanists, ornithologists, entymologists and specialists in almost any branch of natural science the prospect of being able to work in virgin jungle is the equivalent of a child with a sweet tooth being let loose in a chocolate factory. The nineteenth-century explorer Henry Bates perfectly described the forests as 'an uproar of life'. Nowhere else on Earth can one expect to find such a dazzling variety and profusion of animals, insects and plants. In any one acre there are likely to be 300 different species of tree compared with a mere 25 in an equivalent area of British woodland, while a sample zone of 300 square miles will contain anything up to 600 different types of bird – well over ten times the British average.

At the same time that the jungle would satisfy the desire of the scientists for sites which offered the opportunity of original research, it would also provide an ideal proving ground for YE's in search of real adventure in an environment that would test their initiative and their powers of physical fitness and endurance. Because they are inaccessible to any kind of transport – even horseback – the great forests are where you find the few places on the face of this planet which still have not been fully explored. It is true to say that we know a good deal more about the Moon than we do about certain areas of Papua New Guinea, for example, where even the aerial maps have blank spaces because of mist and cloud which permanently hides the terrain from view.

Central to the scientific projects in all three jungle locations was the building of an aerial walkway up to 130 feet up in the forest canopy. That is where jungle life is at its most interesting and yet it is also the one place that the scientists cannot normally get at. Just as the best fruit always tends to be out of reach at the top of the tree so it is with the most fascinating flora and fauna of the forest.

The problems involved in constructing the first walkway in Panama are reflected dramatically in the experiences of young Claire Bertschinger from Hertfordshire, who joined Operation Drake as a nurse and ended up getting involved in some hair-raising high-altitude adventures.

Claire – one of the great personalities to have emerged from Operation Drake – had already experienced a pretty fair taste of the discomfort as well as the excitement of the expedition before she even reached the Caledonia Bay base camp in Panama.

It was 4.00 am on a freezing, foggy December day when I saw her off from Brize Norton along with three other girls. After a twenty-hour flight in a VC 10 of RAF Transport Command they arrived in Belize, where they spent a two-day stopover being royally entertained by British Forces personnel, who were more than delighted to have the opportunity of showing a quartet of pretty girls around the place, before flying on to Panama City. Here they had to use all their feminine charms to get past the immigration officials who were reluctant to let them into the country without return air tickets. Then came the task of finding hotel accommodation at an hour, well after midnight, when four girls coming in off the street tend to be regarded with suspicion. Eventually they managed to find one room between the four of them – two in a double bed, one in a single and one on the floor. The next morning they reported to the local Operation Drake HQ and from there they were sent off for a one-day course at the US Army Jungle Survival School where they were instructed in such gentle arts as skinning a monkey, constructing a shelter out of leaves and branches and milking fresh water from a banana tree.

Three days later Claire and one other of the girls made their way to Colon where at 11.00 pm they boarded a gun boat along with a group of other Operation Drake personnel and set off on the final stage of their journey to Caledonia Bay. Had they been beginning to think by then that it was all a bit of a doddle this illusion was very firmly dispelled in the following few hours.

It was pitch black when they arrived at the quayside and the armed guards who seemed to be everywhere, clearly curious about the presence of two girls in the party, kept shining torches in their faces. At last the silhouette of the gun boat loomed out of the darkness and Claire and the others stumbled blindly down the gangplank. As they advanced cautiously along the greasy deck they noticed that the 65-foot vessel seemed to be in an advanced state of disrepair. The guardrail around the deck consisted of a flimsy wire held up by two foot-high posts, several of which had collapsed. There was very little room aboard and the two girls perched on a shelf near the only lifeboat, which was extremely small and contained only two lifebelts and no life jackets. They were to discover later that there were also no toilets on board and the journey was scheduled to take fourteen hours!

Things didn't seem too bad when the ship first got underway. A cool breeze helped to blow away the overpowering stench of sweat and grime and the fumes of the engine and Claire munched a sandwich. But it was very different once the gun boat cleared the harbour and got out into the open sea. Here the wind was stronger and the water choppy and before long the little craft was pitching and tossing violently among the waves. By 2.00 am Claire and her companion decided that although the prospect of crossing the slippery deck in these conditions was daunting they must try and find a safer place.

After inching her way to the rear of the boat Claire settled for a spot where she could sit with her legs dangling over the edge of the hold and her feet locked around some iron pipes to give her at least some sense of security. She stuck this for a time – preferring to stay in the fresh air rather than go below deck – but the waves kept breaking over the stern and after being hit full in the face several times by sheets of sea water which then ran down her neck she finally gave in and crawled into the luggage hold where she wedged herself between some boxes with a bag full of soggy sandwiches as a pillow. Soon even this was unbearable. People up on deck were being sick and the wind was spraying a revolting mixture of vomit and sea water in every direction. When her boxes suddenly gave way and sent her sprawling Claire decided she would have to go below and retreated to the cabin-cum-engine-room where she collapsed miserably into the arms of a sympathetic warrant officer named Mark.

They clung together in a desperate attempt to keep out the worst of the cold and the wet and, as the sea-sickness tablets she had taken combined with exhaustion to bring on an oblivion of sorts, Claire found herself thinking that death would be a merciful release.

She was brought back to full consciousness by the realisation that the engine had stopped, but it started again after a few minutes and she dropped off once more. When she next awoke the sea was calmer and the sun was shining and when she and Mark rushed up on deck there, lo and behold, was land, a jetty and boats. For a glorious moment she thought they had made it to Caledonia Bay but then the awful truth was revealed – engine trouble had forced them back to Colon!

Claire did make it in the end and after three weeks during which she discovered – and very willingly accepted – that her duties as resident nurse included everything from hairdressing to helping with the clearance of the old, overgrown airstrip she was, much to her delight, sent up to the walkway site to attend to the medical needs of the team there who had quickly discovered that the nature of their work led inevitably to regular cuts, abrasions and other minor injuries.

The spot chosen for the walkway was at a bend in the Turdi River about six miles inland from the main camp. To reach it involved a four-hour march through the jungle which, with a 30 lb pack on your back, was pretty tough going. And wading thigh-deep through murky swamp can be a little nerve-wracking the first few times you have to do it when you haven't developed the confidence to know that there is unlikely to be anything too deadly lurking under the surface of the water. Claire attended to her first emergency when one of her party collapsed with heat exhaustion on the way.

The walkway camp consisted of hammocks slung between trees with a groundsheet rigged up over each one to keep off the rain – quite cosy once you get sufficiently used to the idea of sleeping in what at first seems a very unstable situation so that you are able to relax and stop worrying about capsizing in the middle of the night. The first task had been to clear a helicopter landing pad to enable heavy equipment and supplies to be airlifted in but by the time Claire arrived this had already been completed and work had begun on the walkway itself under the supervision of Royal Engineer Sergeant Mike Christie.

Here the first step was to select a section of forest that

seemed to offer the greatest variety of vegetation and then pick out four suitable trees, each about 150 feet apart, from which to suspend the three connecting walkway sections. Stage one of the construction process involved stringing tensioned polypropelene rope between the trees. Next the framework of the floor, made from Dexion angled metal strips, braced every few inches with cross-pieces, was attached to the ropes. Finally the floor of 16-inch wide expanded metal was added. At its lowest point the walkway was 75 feet high while the observation platforms between the spans were a dizzying 130 feet above the ground.

The trickiest part of the whole operation was climbing the trees in the first place to secure the ropes and set up the pulley system whereby everything and everybody was hoisted up and down thereafter. This was undertaken by Mike Christie, who won the admiration of all and sundry with his agility and ice-cool nerve. The 100-feet ascent, using pitons and spiked boots, would have been quite frightening enough without the hornets' nests and vicious black scorpions that he regularly encountered on the way up. He was actually stung by a scorpion on one occasion but Claire – who had already had to insert five stitches in his hand after he slashed open his palm on a sharp piece of metal – applied a tourniquet, administered an antidote and all was well.

Claire's contribution to the actual construction work consisted at first of being the site 'go-fer' – fetching and carrying materials and sending them up in the hoist. The only danger there came from the nuts, bolts and screw drivers that rained down from above burying themselves inches deep in the earth. But by the time work started on the third and final span several of the team had left and as she was the only helper available Mike asked her if she would be prepared to lend him a hand to fix the block for the main rope 120 feet up. She hesitated – but only until Mike explained that it *was* a bit difficult as it involved having to balance on the branches and added that he would quite understand if she was too scared. Being the kind of girl she is Claire immediately agreed to give it a try. It wasn't long before she found herself deeply regretting this decision.

Going up was all right – in fact, it was rather exhilarating as she was winched up through the branches, past one of the hornets' nests and on up to where the vines dangled down. She wedged herself into the angle between the trunk and a large branch, clipped on what looked like a very flimsy safety line and for the next ninety minutes took the strain of the hoist rope while Mike worked above her head. All was well until the time came to go down again. Mike went first but when Claire followed him the hoist jammed solid when she was still 25 feet from the ground and she had no option but to go all the way back up again to free the knot which was obstructing the block. Once at the top she had to take her weight off the rope while she freed it and this meant half sitting, half kneeling on a branch. As she grabbed for a stray line she momentarily relaxed her grip on the hoist which immediately plummeted to earth before jamming once more – this time so violently that the rope jumped out of the block 10 feet above her head. Now she was stranded.

She admitted later that her eyes began to fill with tears and she started to shake with fear as she realised her predicament. There she was, perched precariously on a wet, slippery branch 120 feet up with no way of getting down and, to make matters worse, ants, flies and spiders were crawling all over her but she daren't brush them off in case she lost her balance. And then the final horror. As she glanced nervously about her she suddenly noticed that on the branch next to hers a snake was crawling purposefully towards her!

The next few minutes took an age to pass as Mike somehow managed to haul his own weight up on the hoist knowing that it might slip at any moment and send him plunging down. Claire tried to concentrate on the breathtaking view and the woodpeckers, humming birds and brilliantly coloured parrots flying nearby in an effort to calm her nerves, but as the giant tree swayed in the breeze her mouth was dry while her skin was bathed in perspiration. She noticed with alarm that her feet had gone to sleep.

Eventually Mike freed the block, lowered himself and sent

the hoist back up for Claire who thankfully clipped on the harness and launched herself into space. She was still trembling several minutes after reaching the ground – but in typical Bertschinger fashion she was up in the treetops again before the day was out.

For the scientists the walkway proved every bit as useful as we had hoped. Ornithologist Peter Hudson from the Edward Grey Institute of Field Ornithology was able to record twenty times more species of bird in the forest canopy than at ground level while Oxford University botanists David Mabberley and Andrew Sugden were able to study at close quarters many plants that would otherwise have been inaccessible. Entymologists Willie Wint from Oxford and Caroline Ash from Leeds were able to do invaluable research into the vertical zonation of insects and zoologist Cathy McKenzie from Bristol University found that her light traps attracted some interesting bat specimens at this high level. Their observations – particularly those bearing on the process of pollination and the respective roles played by birds, bats and insects – will contribute greatly to our previously rather scant knowledge about life at the top in the forest and could also help to explain the mystery of why tropical jungle won't grow again once it has been cut.

Claire Bertschinger's departure from Caledonia Bay was as dramatic and uncomfortable as her arrival had been. The day before the third and final span of the walkway was due to be finished – the 13th as it happens – she went down with a fever and by the next morning it was so bad that it was decided that she would have to be stretchered out. It is hard enough to push through the jungle even when unencumbered by anything more awkward than a small pack but it becomes a nightmare when you have to manhandle a stretcher through the thick undergrowth, across the rivers and through the swamps. However, Mike Christy, Peter Hudson and two Gurkhas broke all records in getting Claire to the nearest point on the coast where an inflatable was waiting to take her on to the base camp at Caledonia Bay. From there she was flown on to hospital in Panama City.

By the time she arrived there her temperature had risen to 103 degrees and she was highly delirious. During the next few hours she was visited by a succession of doctors who ordered one blood test after another without being able to discover exactly what was wrong with her. It was nearly a month before she was declared fully fit again whereupon the amazing Miss Bertschinger could not be restrained from heading straight back to the walkway site. Furthermore, she was delighted to be told that she could join us again in the jungle in Papua New Guinea and immediately started making plans for raising the money to finance a second stint. Talk about a glutton for punishment! As a tribute to her indomitable spirit and her hard work the walkway was named after her. An even greater thrill was to be singled out for personal congratulations from General Sir John Mogg, who not only visited the walkway but insisted on being hoisted up to try it out for himself.

Meanwhile, Claire was not the only person to suffer from a mystery illness. Capital Radio listeners no doubt had a good laugh when it was announced that their reporter David Briggs had had to be flown out of the jungle suffering from a severe case of what was thought to be Kissing Bug Disease. But it was far from a joke as far as poor old David was concerned. His problems started when he boldly volunteered to join the first of the jungle survival courses. These involved being despatched into the forest with no food and very little water and the blunt order not to come back for a week. In the end David didn't last out quite that long!

During the voyage of the *Eye of the Wind* David had proved to be not only one of the most popular characters on the ship but also the smartest – managing to appear elegant and well groomed at all times, even when fighting to maintain contact with Capital Radio in a gale-swept radio shack with 6 inches of sea water slopping around his ankles. Clearly such a man was not about to compromise his sartorial standards for the jungle. He presented himself on the morning of departure looking like a cross between an Action Man marine commando and Stewart Grainger in one of his big white hunter roles. The rakish angle of the hat, the stylish improvisation of the towelling cravat, the buccaneering look of the well-holstered machete – it all added up to a picture so dashing that the photographer didn't hesitate to capture it for posterity.

David wasn't actually quite as self-assured as he looked. Before setting off he asked the Guardia Nacionale escort who was to accompany them whether he thought he was sufficiently well prepared for the venture. The man looked him up and down, grunted and inquired whether he had thought about snakes, to which David replied that he had thought a great deal about snakes! It was then suggested that one fool-proof way of making sure one wasn't bothered by snakes was to put some garlic inside one's boots. Like most people David had started off by treading very carefully in the jungle, being convinced that something deadly lurked under every leaf, and although he had by this time gained total confidence when it came to the well-worn forest paths he was still a bit apprehensive about the horrors that might be slithering about off the beaten track so he took the advice and stuffed his socks with several cloves.

A torrential tropical downpour ensured that the party – led by Captain Christopher Lawrence and including five YE's and the leader of the Operation Drake film team, Alan Bibby, plus one of his cameramen – were soaked to the skin before they had gone more than 200 yards. The first day's march ended at around 3.00 pm and as David was having the greatest difficulty in rigging up his hammock securely the others left him to get on with it while they went off in search of food to add to the somewhat scrawny owl that had been blasted into oblivion from almost point blank range on the way up. Their parting request to him to have a camp fire going ready for their return proved to be easier said than done and when they trooped back dismally an hour-and-a-half later clutching five very small fish David had only just succeeded in setting light to the damp twigs he had gathered despite the aid of a gas lighter. His humour was not improved when he took his boots off before retiring for the night only to find that the garlic had cut his feet to ribbons.

It was on the second night that he began to be obsessed by food – or, rather, the lack of it. The hunting party had returned to camp with a few more fish and some roots that tasted vaguely like potato – a combination that turned out to be less than totally appetising. He spent the rest of the evening reminiscing about his favourite London restaurants.

A further disappointment awaited him when he went to turn in. Having despaired of his hammock, he had plumped instead for a tent arrangement made from his groundsheet with his sleeping bag resting on a mattress of leaves. The finished construction had looked most inviting but, while he was eating his meagre supper and fantasising about slap-up meals in the Savoy Grill, giant ants had munched through his leaf mattress! The only consolation was that the creatures apparently knocked off work at sundown so he wasn't molested during the night.

A far greater problem was the rash that had developed on his arms and face. He had noticed it on the first day but had dismissed it as some kind of heat rash. On that second day it had got considerably worse and by the next morning had turned to a mass of very unpleasant blisters. What with that and his feet David was in a pretty bad way and when the film crew decided at the end of the third day that they were going to return to camp he thankfully went with them.

The first thing they did once they were out of sight of the others was to eat some of the biscuits that the film team had taken with them on the grounds that they needed to keep their strength up for filming and although they were iron-hard army issue they tasted to David like the most succulent steak.

Despite this nourishment he was still a sorry sight when he eventually limped back out of the jungle and the photographer who had recorded his departure could hardly believe it was the same person. He insisted on taking some shots of the grim-faced, bedraggled figure before David was led off to see the medics. They decided he was a suitable case for treatment in the Panama City hospital where it was eventually concluded that he had not Kissing Bug Disease but a form of jungle eczema. Unfortunately for David's dignity the original diagnosis had already been flashed back to London where the news was greeted with unsympathetic guffaws by his fellow executives at Capital Radio not to mention an audience of millions. Six months later he still had the scars to show that it really wasn't so funny. The only consolation was that when he was discharged from hospital and returned to Caledonia Bay he was attached to the diving team and spent most of the rest of his time snorkling among the coral reefs.

Chapter 4

The Beaver light aircraft shuddered violently in the stormy turbulence and the needle of the altimeter had fallen below the 300 feet mark before we finally broke clear of the cloud and then everyone caught their breath in surprise as they realised how close the treetops were beneath us. Only the pilot remained totally calm as he casually eased back on the controls and levelled off at around 100 feet above the dense green forest canopy. After just a few moments we picked out the long-abandoned and overgrown airstrip that was our main landmark and then eyes strained and cine cameras whirred as we made several passes up and down over the surrounding area and then along the nearby coastline where the vegetation came down almost to the water's edge and where the fringe of beautiful but deadly coral reefs were plainly visible through the crystal clear shallows. I detected certain features vaguely outlined in the undergrowth that roused my explorer's curiosity and then, without warning, the view was completely obscured by a spray of oil which covered the windscreen. This time even the pilot registered mild concern. It seemed that an oil seal on the propeller had failed and that meant heading straight for home. As the little plane banked away I knew that I would be back. In the fifteen minutes we had spent looking down on the remote and historically tragic area of the Panama isthmus known as Caledonia Bay I had seen enough to convince me that the place was crying out for the attention of an expedition. Seven years later the first Operation Drake landing craft hit the beach.

That first aerial recce took place way back in 1972 during

an interlude in the Darien Gap Expedition. We had been brought to a halt by recurring back axle problems with our Range Rovers and rather than sit twiddling my thumbs while this was sorted out I took the opportunity to fly up in the Beaver and take a look at this isolated and inhospitable spot which, over the centuries, had provided the setting for so much high drama.

It was Peter Marett – Information Officer on the Darien Expedition and Assistant Director of Administration for Operation Drake – who first drew my attention to the extraordinary history of the place, in particular the ill-fated attempt to found a Scottish colony there at the turn of the seventeenth and eighteenth centuries. This scheme was the brainchild of a pioneering Scots businessman and visionary named William Paterson whose more lasting achievements included helping to found the Bank of England. And although his daring Darien venture went horribly wrong in the end, the thinking behind it was as brilliantly imaginative as it was bold.

Paterson saw the narrow isthmus separating the Atlantic and Pacific oceans as 'the door of the seas and the key of the Universe' and he dreamed of creating a port that would serve as a gateway between East and West and would control the richest trade in the world. To this end the Company of Scotland was formed and in July 1698 five ships set sail from Leith bound for 'some place or other not inhabited in America ... and not possessed by any European Sovereign, Potentate, Prince or State to be called by the name Caledonia'.

By the time they arrived, four months later, at what seemed a perfect location on the remote Darien coast, their numbers had already been reduced by forty. It must have been a terrible voyage. The heat melted the pitch between the ships' timbers and, even when they were under way, the fetid air below decks was thick to breathe and foul to taste. And once the fearsome fever called flux took hold, the stench of death was everywhere as the wretched victims choked on their own black vomit. Deaths were recorded at a rate of up to three a day and, as panic set in, bodies were heaved overboard as soon as the dreaded symptoms took a real hold without waiting for the final, inevitable demise.

For the survivors who made it to their lonely destination the fond idea that they were over the worst of it was very soon cruelly dispelled. The local Indians proved friendly enough but the climate did not. Once the rainy season set in, bringing with it the mosquitos, yellow fever and malaria started cutting a swathe through the colonists who had anyway been severely weakened – first by the voyage and then by poor and inadequate rations. Within six months one quarter of the population of the settlement lay buried in the cemetery that was by far the fastest-growing feature of the community, with new graves being added daily. Plans to build a town called New Edinburgh were soon abandoned while the construction of the fort, with its star-shaped bastions and palisade wall of earth and stakes and surrounding moat, proceeded slowly because even those men who weren't actually dying were scarcely living but were reduced to hollow-eyed, yellowing skeletons too weak to work.

Morale was already at rock bottom when two items of stunningly bad news from the outside world turned festering despair into outright panic. The first was that the King of England had issued a proclamation barring all contact with the colony, thereby cutting them off from any hope of assistance or re-supply. Then in June 1699 a French ship arrived in the harbour with the even more frightening intelligence that a new Spanish Governor had taken over in Cartagena and was in the process of preparing to mount a massive attack against the Scots.

That was the final straw. There was a rush to board the remaining four ships – one of the original five had been captured by the Spanish when she sailed out in search of fresh provisions – and soon only Paterson, his second-in-command and a handful of men too sick to move were left ashore. At that point even Paterson – who had himself now fallen ill – finally accepted that there was no point in trying to soldier on and allowed himself to be carried aboard the *Unicorn*. The pathetic little fleet then set sail leaving behind six men who were too far gone to care very much what fate awaited them but who preferred to take their chance where they were rather than face the miseries of another voyage in the stinking ships. It probably wasn't a bad choice – their reputations, at least, survived when the others were later vilified as quitters. And of the four ships, the *Unicorn* and the *Caledonia* eventually made it back to New York – but not before they had lost a further 255 men – while the *Endeavour* was abandoned at sea and the *St Andrew* limped into Jamaica with 137 fewer men than she set out with, having been chased by the Spanish. The *Caledonia* was the only vessel to return to Scotland and when she sailed into the Clyde she had less than 300 people aboard – one quarter of the 1,200 who had started out so confidently fifteen months previously. Poor old Paterson was left a broken man – driven literally mad by the shattering of his dream.

But that was not by any means the end of the tragic saga. In August 1699 – before the *Caledonia* arrived back with the whole awful truth – a second expedition had set off to reinforce the colony. Over 1,300 men, women and children crowded onto four vessels and sailed out cheering and waving to join what they imagined would be a thriving and well-established community. They lost 160 people on the way over and one can imagine the horror they felt when they got there. One wrote in his diary, 'Expecting to meet with our friends and countrymen we found nothing but a vast howling wilderness, the colony deserted and gone, their huts all burnt, their fort most part ruined, the ground which they had cleared adjoining the fort all overgrown with weeds'.

The second expedition fared just as badly as the first. Fever was rampant again within a matter of weeks and the gruesome daily task of burying the dead once more became the main activity. To make matters worse, there was now a continual threat of attack by the Spanish. Spirits were raised briefly when Alexander Campbell of Fonab arrived and immediately led a daring and successful raid into an enemy camp, where the Spaniards were gathering in preparation for an assault, but the rejoicing at this brave triumph was shortlived. No sooner had Fonab returned to the settlement than a large Spanish naval force began to assemble off Caledonia Bay. Men were put ashore and the Scots retreated to their fort where the fever continued to do the Spaniards' work for them – killing off the besieged colonists far more quickly than the snipers could ever hope to do. Fonab was all for sticking it out until the last man perished, but after a month the Spanish general made an offer they couldn't refuse – a truce of fourteen days while the Scots loaded their ships and sailed away. And so it was that in April 1700 the great Caledonia Bay venture came to its final inglorious end. A large part of the nation's entire capital had been invested in the scheme which had promised untold riches and yielded absolutely nothing. Within a few years the jungle had swallowed up the few remaining traces of their presence so that it was as if they had never been there at all.

For the next 278 years the area remained largely undisturbed. A Dutch plantation company came and went and left behind the overgrown airstrip which helped us identify the spot that day in 1972. And a few explorers, lured by the colourful and melodramatic history of the place, went along and poked about from time to time. But apart from that the ghosts of those Scots, whose many graves soon disappeared beneath the undergrowth, had the place exclusively to themselves until the first Operation Drake landing craft came ashore and Captain Jim Winter of the Royal Engineers advanced into the mangrove swamp to be greeted in most unfriendly fashion by a striking fer de lance – one of the world's deadliest snakes – which he just managed to behead with his machete before it bit him. Clearly Caledonia Bay had not grown any more hospitable in the intervening years!

I had already, in 1976, led a small reconnaissance party to the area and this had confirmed what the rich history of the place and the brief aerial survey four years before had suggested – that here was an ideal spot for a multi-purpose expedition. One of the interesting features of the landscape I had noted from the air was what looked as if it could be the scar left by a long-overgrown moat and, on the ground, we were quickly able to establish that this was indeed the case. I literally stumbled into the shallow but obviously man-made channel that could only be the outer defences of Fort St

Andrew. This was further backed up by the discovery of broken bits of pottery and also a small cannonball.

What was more exciting – and far more significant from an archaeological point of view – was that an old Indian medicine man took us a little way up the coast to another secluded bay from where he led us into the jungle and, after searching, relocated an ancient section of wall that he had first uncovered years before while clearing a space to plant coconuts. There seemed a good chance that here, at last, was the fabled Lost City of Acla – home of the charismatic Spanish explorer Vasco Nunez de Balboa who, in 1513, became the first European ever to gaze upon the Pacific Ocean. If so, it was a major find. And even if it wasn't, our archaeological team were guaranteed some fascinating detective work.

But that was not all. The reconnaissance also provided a lead to the solution of another mystery that could turn out to be financially, as well as historically, rewarding. According to the records of the Scots colonists, a visiting French merchant ship – the *St Antony* – had sunk in the Bay in 1698 with a fortune in gold and silver aboard. With the help of maps and information in John Prebble's book *The Darien Disaster* I selected a point above the coral reef, just where the ship was supposed to have foundered, for a test dive. I had only snorkel gear and could go down no more than twenty or thirty feet but US Army Sergeant Mel Trafford, who accompanied me, was fully equipped and was able to explore the seabed. And here, sure enough, he soon came across a mound 100 feet long and 15 feet wide with lumps sticking out of it. This seemed to me as if it could well be the hull of a ship buried deep in the sand with encrusted spars projecting. We had no time to investigate further but I was satisfied that here was something that would keep our diving team happy – especially as there was the added incentive of treasure trove! But it wasn't going to be easy pickings. Apart from the difficulties involved in breaking through the wreck's cocoon of sand and coral, there was an extra complication that I had become only too chillingly aware of as I floated above Mel. At one point I turned round and came almost face to face with a hammerhead shark! The ugly brute seemed to give me no more than a cursory glance with its beady, unblinking eyes before veering away with one movement of its great tail and disappearing into the underwater gloom but I made swiftly for the surface and reflected that our divers would have to keep their wits about them.

By the time Jim Winter led the advance party of the expedition ashore, therefore, we had been able to pre-plan three major projects in addition to the jungle walkway in some detail. But before we got started on any of those we first had to clear the small and very basic airstrip and set up our base camp. With everybody pitching in this was soon done and we settled quite comfortably into our tented village next to the bumpy but adequate grass runway. By normal expedition standards this was luxury living. There was even a wooden toilet – a magnificent thunderbox constructed near the airstrip and regarded by its architect with justifiable pride until the downdraught from a Guardia Nacionale helicopter landing nearby blew the thing down and left some poor devil sitting enthroned in splendid, if rather embarrassed, isolation! But apart from such minor inconveniences the set-up was very well-ordered.

It would certainly have been the envy of those poor Scotsmen all those years ago. One of their main problems was the rotten, maggot-ridden food they had to force down most of the time. The worst you could say about Operation Drake's rations was that they tended to get rather monotonous after a time. As for liquid refreshment, we had a veritable EEC-style beer lake as a result of a mistake – probably deliberate – in ordering. It was estimated, with some delight, that even if every man drank eight cans a night we'd still be pushed to get through it all before we left Panama – and yet we actually ran out with six weeks still to go! If nothing else, that just goes to prove that exploring in the tropical heat is thirsty work.

Snakes were what worried people most but although we saw plenty of them nobody got bitten. There were one or two nasty moments. Archaeologist Andrew Hunter was crawling on hands and knees on the site of Fort St Andrew looking for pottery relics when he suddenly found himself looking right into the face of an 8-foot boa. Luckily they are quite harmless and he grabbed it and brought it back to camp. On another occasion I was walking past one of the tents when Mark Moody, who was sitting inside writing up his project notes for the day in his usual methodical way, called out very casually, 'Excuse me, sir, but do you know much about snakes?'

'Yes – a little,' I replied. 'Why?'

'Well, there's one coiled round my foot, sir,' he explained conversationally.

And so there was! It was rather an attractive, peach-coloured thing and it was lying with its head on Mark Moody's bare foot and appeared to be asleep. I told him not to worry, since it looked like one of the non-poisonous varieties, but had to admit that it could be something more dangerous and advised him to sit still until a zoologist could be summoned to make a positive identification. It was eventually confirmed as a boa – but not before poor Mark Moody had had to sit motionless for several minutes surrounded by a circle of curious sightseers who speculated loudly on such matters as whether it was thirty seconds or thirty minutes that one could expect to live after being bitten by a fer de lance.

As it happens, I probably had the most alarming brush with Panamanian wildlife during the 1976 reconnaissance. Having been told that there were no mosquitos around at that time of year I retired to bed without the protection of a net. Just before I fell asleep I felt a slight pinprick in one of my toes but thought nothing of it. However, when I awoke in the middle of the night to answer a call of nature and switched on my torch I suddenly noticed to my horror that the bed was soaked in blood, which was still issuing from my toe at an alarming rate. At the same moment I felt something brush past my face and, swinging the beam of the torch upwards, I was just in time to catch a glimpse of a bat flitting out of the room.

I had been bitten by a vampire bat and the reason there was so much blood about was that these creatures inject you with a form of anti-coagulant to facilitate their loathesome feeding. I recalled with a shudder how our horses had suffered during the

Darien Expedition of 1972 – being drained nightly in the grimmest Dracula tradition until they were almost too weak to stand. I didn't suffer in that respect – although I did lose an estimated half-pint of blood – but the anti-rabies shots I had to have as a precaution were most unpleasant. Altogether I had to have thirty-two jabs in my stomach and backside and apart from being extremely painful they left me dazed like a zombie. The only consolation was that since I had enjoyed several large scotches before retiring that night the bat probably died of alcoholic poisoning!

In charge of the archaeological projects at Caledonia Bay was a great young character from Cambridge University called Mark Horton. Despite being only twenty-three, Mark struck me immediately as being the Magnus Pyke of archaeology. He has the same air about him, he waves his arms about in the same way and he has the same boundless enthusiasm for his subject. He would get so involved in telling you about his discoveries at the various sites that he would be quite oblivious to what was going on around him. On one famous occasion when General Sir John Mogg visited us and Mark was giving him a guided tour of Acla, a snake slithered across the path right in front of the general, whereupon I reached for my revolver only to be beaten to the draw by a Guardia Nacionale major who had a shotgun. We both blasted the snake and, as the sound of gunfire died and the dust cleared and people almost literally picked themselves up off the floor, Mark was heard to be prattling on about his beloved artefacts as if nothing had happened.

His great sense of humour stood him in good stead in the face of an endless succession of practical jokes and boisterous pranks played on him by frustrated YE's who didn't all share his enthusiasm for crawling around in the sand and the coral dust under a blistering sun searching for fragments of 17th century Scottish pottery while at the same time trying to dodge falling coconuts. He grew accustomed to being frog-marched to the beach and hurled fully-clothed into the water, tied down over ant hills and generally tormented and took it all in good spirit.

In between the horseplay he did a marvellous job for us and must personally take much of the credit for the great success of the archaeological projects. The layout of Fort St Andrew was precisely plotted and a mountain of fascinating finds were unearthed. These included bronze buckles, spurs and a Scottish coin bearing the crown and thistle and the date 1695 as well as pottery, beads for trading with the Indians and a host of clay pipes. Excavation of the three-sided South bastion of the fort's ramparts resulted in the discovery of a hoard of military relics including musket balls, flintlocks, pike ends, knife blades and a bronze rapier and part of another sword.

Even more thrilling were the discoveries that enabled Mark to confirm that the spot to which I had been led by the old Indian medicine man was indeed the site of the Lost City of Acla. It was never actually more than a village but it gained great historical significance as the place where the amazing Balboa made his home, having married the daughter of the local Indian chief, and from where he set out on his epic trek across the isthmus that ended with him becoming the first European to look upon the Pacific – an achievement sometimes wrongly attributed to Hernando Cortes. It was also the place where he died – beheaded after falling out with the man who succeeded him as Governor of the area.

With the death of her founder, Acla seemed to just disappear. Later, historians began to speculate about the exact whereabouts of the place. But although they were able to establish that it was somewhere along a fairly short stretch of coast near Caledonia Bay, that it had a natural harbour and that it was between two rivers, nobody was able to use this fairly precise information to pinpoint the location. With the wisdom of hindsight we could understand why this was. In the first place, it looks from the sea as if the coast at this point, far from providing a harbour, is actually protected by an impenetrable reef. It is only when you get very close that you realise there is a way in to what then amounts to a lagoon about 100 yards across. Secondly, sandbars conceal the rivers. And finally, a Spanish fort was built on, or near, the site some 200 years later and this became something of a red herring. But

Mark Horton's very thorough investigation of the spot resulted in the discovery of pottery and other relics which, when dated, left no doubt that this was indeed Acla.

Meanwhile, the diving team had an initial set-back when the mound which had excited Mel Trafford and I so much during the 1976 recce turned out to be not the *St Antony*, with its precious cargo of 60,000 gold ducats, but merely a natural feature of the seabed. However, this disappointment was more than made up for by the discovery of another wreck which proved to be just as important historically as the *St Antony*.

Wreck detection expert Tony Lonsdale had been patiently checking every square yard of the bay with his proton magnetometer but all his more interesting readings turned out to be false alarms until he moved in to the sheltered area on the eastern side which seemed the likeliest spot for the colonists to have had their anchorage. Here he picked up something really promising and exploratory dives ccnfirmed the presence of large sections of planking buried in thick silt and coral at a depth of about 20 feet. They were definitely on to something.

Underwater visibility in this area was particularly bad – especially once the silt was stirred up – but gradually more and more of the timbers were uncovered. Eventually a loose section was brought to the surface for close examination by marine archaeologist Meredith Sassoon. Her verdict sent a thrill of anticipation through the team – for the timber was an unusual sandwich of oak and pine with a layer of fibrous tar in between such as was known to have been used in old sailing ships. The next specimen to be raised caused even more excitement. It was clearly a piece of deck planking and it was badly burnt on one side. Also, the nail holes in it were charred right the way through which indicated that the fire had been so intense as to cause the nails to glow red hot. The significance of this was that it pointed very definitely towards one particular ship – the *Olive Branch*.

The *Olive Branch* was one of the ships that brought the second wave of colonists from Scotland in 1699. One night a drunken crew member sneaked into the hold intent on tapping one of the barrels of brandy that were stowed there, but, in his befuddled state, succeeded only in dousing himself and the lower deck with the raw spirit. He then panicked, dropped the lantern he was carrying and turned the vessel into a blazing inferno. Within a matter of minutes the ship had been burned to the waterline and had sunk with all her stores still intact. According to records still in existence these supplies would have included Scots bonnets, Bibles, buttons, beads, glass drinking cups, hunting horns, mattocks, looking glasses, pewter jugs ... and a quantity of clay pipes.

It was this last item that enabled us to clinch the identification of the wreck as the *Olive Branch* when, after three weeks of careful excavation, a number of barrels were uncovered in what must have been the cargo hold. On the seabed the divers carefully exhumed the first five barrels – still largely intact – and delicately sifted through the mud and silt which filled them. Aboard the 'David Gestetner' – the giant inflatable that served as a diving platform – there was an air of breathless anticipation as the finds panned like gold in this way were passed up for examination. The first one yielded a two-pronged fork and a few slivers of bone – evidence, perhaps, of a consignment of salt pork or beef? The second and third were both empty. The fourth was full of ferous nails. And in the fifth were found ... three clay pipes in perfect condition! Any lingering doubts that it was indeed the *Olive Branch* that lay below us were removed when Mark Horton confirmed that the pipes were exactly the same as those he had found on the site of Fort St Andrew. They even had the same initials – P and G – stamped on them. The celebrations that followed considerably reduced our beer lake! To round things off perfectly, the discovery of the barrels coincided with the visit to Caledonia Bay of Operation Drake chairman, General Sir John Mogg.

It also helped to boost the confidence of the diving team whose enthusiasm for the project had slumped dramatically a few days before after a nerve-wracking incident. Once again, it was Mark Moody – the man who had the snake coiled round his foot – who was at the centre of it. He was down on the seabed working in zero visibility when the bucket he was filling with the silt he cleared was suddenly snatched from his grasp. He assumed that those above must have hauled it in – until he surfaced to find them asking him what *HE* had done with it! Its disappearance remained a mystery until a chilling possible solution was provided a few days later when the Guardia Nacionale hauled in the fishing nets they had set nearby. There, threshing in the mesh, were a number of sharks including two man-eaters – a 5-foot Mako and a 10-foot Hammerhead. Thereafter the sensation of having a fellow diver brush against one accidentally while groping blindly in the gloomy depths was enough to make the coolest customer jump right out of his wet suit.

While all these dramas were being enacted at the various project sites, life at the base camp had settled into a reasonably well-ordered routine disturbed only by fairly minor crises and the visits of VIP's and journalists from all over the world.

It was the arrival on the scene of a writer and photographer from a certain Swedish magazine that caused one of the more amusing problems I had to deal with. They came down from New York after first contacting me to get my permission for a series of feature articles. Although I hadn't heard of their publication the circulation figures they quoted were impressive and I was delighted at the prospect of more valuable publicity. But no sooner had they arrived and trotted off to start getting their interviews and pictures than I was notified that they were involved in an angry scene down on the beach. I immediately raced down there to find out what all the fuss was about and the first thing I saw was our attractive lady marine archaeologist Meredith Sassoon fuming with indignation.

Was I aware, she demanded furiously, that these so-called gentlemen of the press were, in fact, representatives of one of the most disgusting hard-core porn magazines ever published? And would I mind making it absolutely clear that no way was she going to pose for pictures clutching her artifacts!

When I inquired innocently whether the magazine was like Playboy she snorted that it was far worse than that – it made Playboy look like a parish magazine! Even the photographer himself admitted that his wife wouldn't have the publication in the house because it was so risqué – but he quickly added that the Operation Drake article would be absolutely straight. In the end I managed to smooth things over and they were able to get the material they needed. It wasn't until some months later that I was sent the published articles. They were spread over four issues and were done very well indeed – intelligently written and beautifully illustrated. But as for the rest of the contents! I could hardly believe my eyes. Some of the pictures would have brought a blush to the cheeks of a naval stoker.

Some other small difficulties arose from a deeply ingrained hostility between the local Cuna indians and the Guardia Nacionale of whom forty officers and men were attached to us as a kind of permanent chaperone force while we were in Panama. The Guardia are the police force and the army all rolled into one under the command of General Omar Torijos – a tough and charismatic figure who, when he visited the camp, shinned up a coconut tree with the greatest ease to fetch my daughter a fresh coconut. I had first made contact with him during the Darien Gap expedition and had very quickly realised that here was a man you needed to have on your side. He runs an all-powerful 10,000-strong militia which operates with ruthless efficiency. Nobody messes with the Guardia and, as we went to some lengths to explain to the YE's, if there's any sort of trouble and the Guardia are about – put your hands up! At the same time, it has to be said that they did everything they could to help us while we were in the country.

The antagonism between them and the Cuna went back many years to the time when there was an Indian uprising during which there were a number of very bloody clashes. We were aware of a simmering animosity and this wasn't helped by the unfortunate matter of the banana and coconut trees. A number of these were accidentally either damaged or totally demolished by our people and after complaints from the Cuna we fixed a compensation price. This was fine until one of the Indian farmers tried it on and put in a claim for trees that hadn't been touched. Unfortunately for him, the Guardia found out what he was up to and exacted a subtle punishment. The farmer was made to sit and watch while his Chief did all the Guardia's washing up and swept out their camp. The Chief was so seething with fury by the time they let him go that he jumped into his canoe and paddled off leaving the farmer to sit pathetically on the edge of the camp for two days before he was finally picked up.

This certainly did the trick and we had no more phoney claims from then on – but it did nothing to foster a spirit of brotherhood and goodwill between the Cuna and the Guardia and, more important, undermined the relationship between the Cuna and Operation Drake. In an effort to repair some of the damage we decided to hold a Grand Sports Festival in which the games were carefully chosen to ensure that everybody would win something. For instance, the Guardia spent their entire time playing volley ball so this was included and they duly wiped the floor with the rest of us. On the other hand, it was patently obvious that nobody could live with the Cuna when it came to canoeing so that was nominated as another event. And for Operation Drake? What else but bowls! At the end of it all, everybody seemed to be on speaking terms again – but whether this was due to the athletics or the drinking and dancing that was laid on afterwards is a matter for debate.

The Cuna are generally delightful people and they were fascinated by our activities at Caledonia Bay. They were often able to help in a very practical way – as did the old man who led us to Acla. There is no doubt that certain events connected with the Scots colony have passed into folklore and we even heard that one could sometimes come across effigies of William Paterson complete with his top hat. That, I suspect, is just as much a myth as the story about a Scottish tribe descended from the handful of colonists who disappeared into the jungle.

I first heard this story in New York just before I set out on the 1976 recce. I bumped into a wild Welsh writer and explorer called Tristan Jones – one of those incredible characters who seem to have been everywhere and done everything. We got to reminiscing about some of the more unusual spots on Earth that we had visited and then I mentioned that I was bound for Caledonia Bay and added,

'That's one place you've never been to, I'll bet.'

'So happens I was there just last week, bach,' he retorted with a note of triumph in his voice. 'And I'll tell you what, boyo, a very weird place it is too.' He went on to tell me that he had actually met a tribe of Indians that included white men with names like Robinson and Farrell. 'I tell you, it's a Lost Tribe of Scotsmen,' he insisted. 'How about that?'

I suspected he was pulling my leg but he swore blind that he was telling the truth. And so he was. Imagine my open-mouthed amazement when virtually the first local I met in Caledonia Bay was a white man with reddish hair and blue eyes, dressed in jeans and a T-shirt, who introduced himself as Robinson. His blonde-haired wife was called Mary.

For a moment I was convinced that this really was undeniable evidence of a link with the Scots colonists of New Edinburgh. But after talking to the couple I soon realised that there was a more mundane explanation for their appearance which contrasted so strongly with the dark, swarthy looks of their fellows. They were albinos. It turned out that there were quite a number of them in the area – probably as a result of in-breeding. The names seemed to be pure coincidence although I like to think that it was not impossible that they had been handed down through the generations, having first been adopted by the Indians following contact with the Scots.

The Scottish connections with Caledonia Bay were used as the excuse for a mammoth Burns Night party which was fuelled with almost unlimited supplies of whisky provided by sponsors J & B. People flew in from everywhere and our little runway was choc-a-bloc with the aircraft of the VIP's. As the night wore on the scenes grew more and more abandoned. Guardia officers twirled their moustaches and pursued nubile YE's along the beach, people ran off with each other's wives, the Panamanians danced around singing 'Viva Grande Bretagne' and cursing the Americans and I ended up sitting on the shore greeting the dawn along with two very high-ranking US naval officers who had fought valiantly to stay sober all night but had eventually been forced to admit defeat.

Whether it is true, as reported, that one guest was witnessed weaving among the assembled company demanding to be introduced to the host, Robbie Burns, so that he could thank him personally for a great thrash, I don't know. But one thing is certain. Nobody who was there will ever forget the sight and sound of Piper Robert Little, one of the Scots Guards, standing silhouetted against the palm-fringed headland as he played a lament in memory of his ill-fated countrymen of so long ago. It was a haunting moment which the ghosts of Fort St Andrew would have appreciated.

Chapter 5

Somewhere amongst the mountain of documentation that charts in every detail the full history of Operation Drake there is a tiny, torn-off scrap of pink paper on which is scribbled a list of four names together with some figures. The filing system in room 5B isn't the most sophisticated and it would probably take hours to sort it out from among the schedules, the reports, the diaries, the memos, the press announcements, the letters, the invoices, the background surveys, the briefings and all the rest of the welter of data and information that we have gradually accumulated over the years. And yet it pinpoints what was, unquestionably, the most dramatic moment of the whole expedition.

The names were those of four members of the Balboa Patrol who had fallen ill during the 100-mile trek that began as an afterthought and then developed quickly into one of the most challenging and punishing projects we mounted. I jotted them down along with their respective temperatures and respiratory rates as they were given to me over the radio from a clearing deep in the Panamanian jungle by a tense and anxious George Thurstan – the patrol leader who had finally put out an SOS when he realised that, in one case at least, the emergency had reached the life-or-death stage. All four men were in a pretty bad way through a combination of dehydration and heat exhaustion, but film cameraman Rik Gustavsen was so ill that if he were not airlifted out immediately there was a real danger that he would not survive.

Had it just been a matter of a helicopter flying in and picking the sick men up this would not have presented much of a problem. What added an extra dimension of difficulty to the situation was the fact that the Patrol, through no fault of their own, did not know exactly where they were – and trying to locate a small party of people somewhere within hundreds of square miles of jungle is almost as tricky as it is to spot a life raft in a similar area of ocean. From the air the forest appears as an unbroken carpet of green stretching in all directions as far as the eye can see. Man-made smoke signals are hard to pick out among the forest fires that smoulder away all over the place and even flares and marker balloons are not easily spotted unless the aircraft is right on top of them or the people inside happen to be looking in exactly the right direction. Back at headquarters in Caledonia Bay we knew as we raised the alert, and the helicopters scrambled, that the next few hours were likely to be very tense.

It had all started seventeen days previously when George Thurstan led his team of twenty-two out of Acla, bound for Santa Fe along the Balboa Trail – the route across the isthmus first blazed through the jungle by the Spaniard in 1513. This venture hadn't been included in our original plans for the Panama phase but was added at the last minute when the Membrillo River project fell through. The Membrillo River is notorious as the home of the Buffalo Gnat – an insect so unpleasant that even the local Indians can't stand it and have moved out of the area. Our aim was to locate this fearsome creature and study it, but certain political problems arose and we had to cancel the whole thing. We were left with just seven days to think up an alternative to fill the vacuum and it was after poring over an historical map of Panama that George and joint phase leader Major Alan Westcob came up with the idea of the Balboa Patrol.

The Guardia Nacionale were not at all keen on it – their first reaction was to rule it out altogether. This was understandable. Their main concern was to make sure that nothing terrible happened to us while we were in Panama and there was no doubt that what we were contemplating here was very risky. Previous attempts had not only failed – they had ended in disaster. A century previously, a joint British, French and

bove: Young Explorer Wizz Gambier holds 'Monty Python' (CHRIS SAINSBURY)
elow: The launch, 1978: Prince Charles at the wheel of the *Eye of the Wind* (CHRIS SAINSBURY)

Above: Overhauling the buntlines halfway across the Pacific (CHRIS SAINSBURY)
Below: View from the aerial walkway constructed in Panama (CHRIS SAINSBURY)

Above: Carrying the inflatable down into the volcano crater in Souffrire (CHRIS SAINSBURY)
Below: Ben Gaskell examining a bat in Papua (CHRIS SAINSBURY)

Above: Andrew Mitchell meets a wild Wana tribesman (RUPERT RIDGEWAY)
Below: Wreckage of a war plane discovered in Papua New Guinea (OPERATION DRAKE)

Me with a 6ft 1½ in lizard in Papua (MICHAEL CABLE)

A Papua New Guinea Wigman (JEROME MONTAGUE)

ove: Keith Crawford leads the way up the Ranu river (RUPERT RIDGEWAY)

low: Roger Chapman, Yog i Thami and Jerome Montague battle with the Upper Strickland
LL NIEUMEISTER)

Above: The *Eye of the Wind* meets *British Resolution* supertanker (CHRIS SAINSBURY)
Below: Robbie Williamson trains Young Explorers for Sulawesi diving project (RUPERT RIDGEWAY)

American expedition had lost several men. The Guardia could well do without a repeat of that kind of tragedy and the bad publicity it might bring. We pointed out that we would be going in the dry season when conditions were not quite as hazardous as they were when the rains had swollen the rivers into raging torrents, that we had a very good back-up organisation and that we weren't about to take any silly chances.

At this point they reluctantly agreed but showed how seriously they regarded the whole thing by flying in some of their very top officials to advise us. These included the Commanding Officer of their equivalent of the SAS, the Commandant of their Recruit Training School and their No 1 signals officer.

We had no illusions about what we were letting ourselves in for. Even in the dry season, with a back up force of helicopters and with the latest and most sophisticated radio equipment, it was going to be very tough – much tougher, surprisingly, than it would have been for Balboa. For one thing, he went with a party of over 500 Spanish and Indian bearers and guides and secondly, the Panamanian jungle was actually a much more civilised place back in the sixteenth century. On the face of it that sounds ridiculous – but it's a fact. There used to be far more Indians in the interior in those days, more villages and a lot more well-worn trails. Later the Cuna and the Choco fought each other and the Cuna withdrew to the coastal areas leaving the forest to close in again. We knew that the patrol could expect to be hacking their way through dense undergrowth much of the time and that they would probably go for days on end without seeing another living soul.

Despite this we had no real reservations about going ahead with it. After all, Prince Charles himself had talked about 'the challenges of war in peacetime' and George Thurstan in particular felt very strongly that this was precisely what the Young Explorers were looking for and that we should not be afraid to stretch them and spice the programme we arranged for them with an element of real danger. George is a tough, no-nonsense ex-policeman who believes in driving himself to the very limits so, coming from him, this view was predictable. But it was also borne out by the comments of the YE's in the final reports we asked all of them to write up at the end of each phase. The most common complaint that was raised was that we didn't make things hard enough for them a lot of the time. Nobody who was on the Balboa Patrol mentioned that!

George picked his team very carefully. Everybody had to undergo an exhaustive medical test in which everything was checked from their teeth to their blood pressure. Several people were turned down because they were not perfectly fit. The party that eventually set out from Acla on 25 February was made up of twelve YE's – including three girls – plus five Guardia, a Choco Indian guide and hunter with the delightful name of Jesus Zuleta, film cameraman Rik Gustavsen, radio operator Alex Gill and photographer and diarist Desmond Dugan. They carried with them clothing, rations and medical supplies that had been pared to the minimum safety requirement after long and careful group debate about priorities.

With everyone fresh and in high spirits the first day's march had the easy-going atmosphere of a Sunday afternoon nature ramble despite the fact that at a very early stage they had to leave the beaten track and push into the thick of the forest. They had time to study with interest some tapir tracks and the Pit Viper which Guardia Sergeant Juan Diaz spotted and pointed out as being the most venomous of all Panama's snakes as well as one of the smallest.

Mid-afternoon is the time one starts looking for a suitable camp site when on this kind of expedition and at about 3.00 pm they found an ideal spot near a small stream. After a meal of dehydrated rations everyone was in a relaxed mood and the camp fire gossip went on late into the evening, while Jesus quietly disappeared with his shotgun and returned sometime later with two agouti.

This augured well for the rest of the trip since the team had set out with only basic rations and the aim of living off the land as far as possible. Jesus proved an efficient hunter and kept the pot well-supplied with jungle turkey, pig, the occasional monkey – and even snake. There's a lot of meat on a 6-foot python! It was interesting to observe how people's attitudes towards this kind of food gradually changed. For some it was, at first, a bit *too* fresh and they found difficulty eating something they had seen slaughtered and butchered before their very eyes. The next stage was that common-or-garden turkey was all right but pretty little monkeys or exotic and beautiful parrots or toucans were not fair game. It wasn't too long before hunger removed all hang-ups. Only Rik Gustavsen held out – and that was because he is a confirmed vegetarian. George noted this with slight alarm. He has little patience with such fads at the best of times and in retrospect he was to blame Rik's collapse partly on his vegetarianism and also his insistence on eating out of a hollowed-out coconut shell, which resulted in him having half rations all the time since the mess tins which everybody else used were twice the size.

Day two had the patrol actually walking in a small river. Although it wasn't very comfortable to splosh along in the water for long periods and although they had to concentrate hard to avoid slipping and stumbling on the uneven bed of the river, it was still far easier than hacking their way through the heavy, tangled undergrowth. During the day the party also had its first taste of two problems that were to arise time and again before the expedition was over. The first was that they got lost. This was because of the near impossibility of plotting an exact position with maps that are inevitably less than totally accurate and in an area where the terrain is not only such that landmarks are difficult to pick out but can also change considerably depending on what season it is. They were only slightly off course and fairly quickly managed to put themselves right – but not before the second in-built complication emerged. At one point there was disagreement between George and Juan Diaz as to which direction they should be heading in and when George put his foot down and insisted on going his way, Juan went off in a huff. It was clear that he considered himself to be joint leader with George and there were to be several more clashes before the patrol ended.

George sensibly adopted a policy of giving way on the small points but digging his heels in when it mattered. As he always proved to be right he had the full support of the team.

The next few days were largely uneventful, although the novelty began to wear off and heads went down as the pace hotted up and the cumulative effect of long, hard spells on the march in testing conditions began to sap reserves of energy. People flopped exhausted into their hammocks in the evenings and the camp-fire chat sessions were growing shorter and shorter.

The unavoidable minor irritations also began to set in – the sweat rashes and insect bites, the blisters and the scratches that refuse to heal in the tropical humidity. Young Scots Guardsman Jock McGregor developed what looked ominously like impetigo and Guardia medic Daniel Rodriguez was having a lot of trouble with his leg. Apart from that there was the mild annoyance of repeatedly straying off course and the excitement of spotting the tracks of some big cats – Jesus reckoned they were either puma or jaguar. There was also a moment of alarm when the powerful Clansman radio started to pack up, but Alex Gill got a message to HQ requesting a replacement on the re-supply helicopter which was scheduled to rendezvous with the patrol at the village of Surcurti on 3 March – day seven.

It was now that the first major crisis developed. The village of Surcurti on the Surcurti River had been picked out as the ideal spot for the re-supply drop during an aerial recce shortly before the patrol set out. George will admit that when he first looked down from the air onto the area he was about to try and cross he did experience just a momentary flash of doubt. He described it as being like flying over the prairie wheat lands except that, instead of corn, it was a sea of green trees that stretched unbroken to the horizon. After a while the pilot identified the Surcurti River, along which the patrol's route lay, and having followed it for some miles, came upon a large village which he confirmed as Surcurti. It was close to a bend in the river where a broad gravel bank formed what would serve as a perfect landing pad.

Major Eric Aguillera – the officer in charge of the Guardia

unit that had been attached to Operation Drake in Panama – was with George and Alan Westcob in the plane and he made a point of stressing that whatever happened the patrol must be sure that it went to Surcurti and not to nearby Morti. This was on the Morti River which was very similar to the Surcurti. He was vague about his reasons for emphasising so strongly the need to avoid Morti but later that evening, over a few scotches, George wheedled from him the somewhat startling information that the villagers there had once captured a Guardia patrol and held them hostage and it had taken a major operation to go in and free them.

The Balboa Patrol picked up the Surcurti on the afternoon of the fifth day and immediately pitched camp. They got going again earlier than usual the next morning so as to make sure that they reached the village and got settled in with plenty of time to spare. Eventually they arrived at a bend in the river around the corner of which, according to the map, lay Surcurti. This seemed to be confirmed by the presence on one bank of a banana plantation – although this appeared to have been neglected so that the fruit was rotting on the trees.

George called a halt and, having warned people not to take pictures of the Indians in case it frightened them, led an advance party forward round the river bend. The sight that met their eyes was totally unexpected. There was nothing. Obviously there had once been a village there but it had long been deserted.

Morale immediately plummeted. Everybody had been looking forward to a couple of days resting up in the relative luxury of a village environment and the boost of getting fresh supplies. There was also the worrying question as to what had gone wrong. For once they were not in any doubt about their position. They were definitely on the Surcurti and the abandoned village was certainly the one they had been aiming for. So what to do next? To make matters worse, the radio had by now ceased to function altogether so there was no way of contacting headquarters to let them know what was happening and to seek advice.

Again, there was disagreement between George and Juan

Diaz but in the end George prevailed and the party set off up the river for the next village in the hope that this was where the confusion had arisen. They camped that night at a spot where a chopper could have got in and, their spirits raised once more after a good meal of American rations supplemented with fresh-picked bananas, they decided that the next morning they would hoist a marker balloon and hope that the re-supply aircraft would spot it.

They got up early and sat for some time straining their ears for the unmistakable thud of helicopter blades – but in vain. Eventually they accepted that they must move on and they left the marker balloon plus a giant arrow of leaves pointing in the direction they were travelling. Progress was slow because they had to leave the river and take to the jungle, where the undergrowth was so dense that visibility was down to a few yards and by the time they had to stop for the day and make camp they had still not reached the second village.

They were further delayed the next morning when they awoke to find that Jesus and Sergeant Arsimodo Jaen of the Guardia had failed to return from the previous evening's hunting trip. They walked in minutes after a search party had left to look for them, and explained that they had got lost whilst chasing a pig and had decided to wait until daylight before trying to find their bearings again.

The second village was reached after a couple of hours but it was obvious before they got there that it was going to be just as dead as the first one. However, no sooner had they arrived than a helicopter flew over and although it was very high they managed to attract its attention with a smoke flare. There was nowhere suitable for it to land but the pilot, giving a brilliant exhibition of controlled flying, slotted it in above the river bank with its blades clipping the trees and hovered there long enough to unload most of the supplies. He then flew off to search for somewhere to put down and found a clearing not too far away.

The mystery of where the carefully-planned rendezvous had gone wrong was now cleared up. The village which the pilot of the recce flight had identified as Surcurti was actually the very

Morti which George had been warned to steer clear of. The confusion arose because the pilot had picked up the wrong river – a mistake easily made. The re-supply helicopter duly landed there and the pilot only realised that there had been a mistake in the nick of time and took off just as the hostile villagers were advancing menacingly towards him!

The patrol stayed put that night and the next morning the helicopter returned with a new radio. There followed another lazy day during which everybody took the opportunity to wash themselves and their clothes and generally take stock of the situation. A number of people had started suffering from foot rot – which had probably been picked up at the Choco hunting camp site where the party had spent the night two days previously – and others had gone down with dysentery. George Thurstan had a particularly bad dose and was feeling so bad that there was even a suggestion that he shouldn't go on for a day or two until he improved, but, characteristically, he pooh-poohed any such notion. Less debilitating, but equally unpleasant, were the ticks with which nearly everyone was now infested. The patrol sometimes resembled a tribe of baboons as they helped to rid each other of these loathesome lice that resisted removal like limpets and produced the most awful sensation as they crawled up between the hairs on one's legs during the night. The daily tick count became quite an event as people vied for the record of having the most embedded on his or her body and at one point the average per person was nearly forty!

A more delightful entertainment was provided that day by a tree snake which passed through the camp. It cruised through the branches at an impressive speed and never once lost its smooth rhythm as it moved from one tree to the next. Attempts to catch it were futile and it soon disappeared purposefully on its way at the same steady rate.

On 6 March – day ten – the patrol got moving again and pressed on to the Chucunaque River of which the Surcurti is a tributary. They reached it the following day without any major incident although there was a momentary panic when, without any warning, two of the Guardia men at the rear of the column opened up with a barrage of rifle and shotgun fire apparently aimed at nothing in particular. Everybody stopped and gaped in amazement as they blasted away at the water and then all became clear as an ugly serpentine head broke the surface only to be vaporised almost immediately by a final volley from the shotgun. It was a fer de lance – nicknamed Mr X because of the markings along his back. Deadly, sure – but there were those present who felt that they had been in marginally more danger from the fire of the Guardia.

For a couple of days the patrol followed the course of the Chucunaque and it was here that they encountered some of the most impenetrable jungle of the entire trek. Progress at one point was reduced to less than two miles in a whole morning. Then, after a final visit from the re-supply helicopter when everyone was checked by a Panamanian doctor and pronounced as well as could be expected, they headed away from the river on what was expected to be the last lap of the slog to Santa Fe.

They anticipated that their problem from here on in was going to be a lack of water – and so it proved. At first it was a novelty to be without it, having been for so long either very close to, or actually wading through, an abundance of the stuff. It was a matter for jokes when they had to start rationing themselves as the bottles they had filled before they left the river began to empty. They were still fairly lighthearted about it even when they had to start filtering the contents of stagnant pools through Millbank bags to get rid of the solid impurities before boiling it. But it wasn't too long before the novelty wore off. It very quickly got to the point where George Thurstan had to alert HQ to the possibility that another urgent re-supply flight might be necessary as even the brackish puddles were found to be dried up. The next morning Rik Gustavsen woke up with a fever.

They broke camp two hours earlier than normal that day and moved out without any breakfast at 6.30 am with the intention of covering as much ground as possible and thereby increasing their chances of finding water. But it soon became clear that they now had another problem. Since the previous day they had been following an old timber trail which was supposed to take them on to the Pan American Highway just short of Sante Fe, but it had been gradually leading them further and further in the wrong direction. Juan Diaz and the Guardia kept insisting that, if they persevered, the trail would turn eventually and put them back on the right bearing. But every bend seemed to take them even more out of their way. George Thurstan was getting anxious – yet there was nothing he could really do about it. With such an acute shortage of water, the alternative of plunging back into the thick of the jungle and hacking their way along on course was out of the question. Not only would the extra effort dry them out even quicker, but their progress would be so slow that they would greatly reduce their chances of finding water.

Meanwhile Rik was rapidly getting worse. The only way to keep him going was to share out his kit amongst everyone else and as he grew weaker and weaker he soon had to be half-carried. By the time a halt was called in the early afternoon the whole party was mentally and physically exhausted. Then came the last straw. As they sat resting and trying to resist the temptation to gulp down the half cup of water that was all each of them had left, the Guardia para-medic, Silvester, suddenly keeled over with giddiness and vomiting.

Now that he had two sick men on his hands George Thurstan no longer had any choice but to call in a rescue helicopter. Even so, there was no point in raising the alarm until a landing site had been cleared.

Despite being almost totally drained, the patrol somehow managed to galvanise themselves into action and get stuck into this seemingly impossible task. Using nothing but their machetes, they felled trees up to two feet thick and within four hours had cleared an area roughly 50 yards square. In the circumstances it was an incredible feat – but it inevitably took its toll. Two YE's collapsed with heat exhaustion – David French from England and Irishman Miles Clark, who had kept everyone amused with his endless fantasising about food and who was found to have lost 25 lbs when they eventually got him back to base. It could have been worse. At one point a tree toppled unexpectedly and when Cathy Davies turned to run out of the way as the 'Timber!' cry went up, she caught her foot in some of the loose branches that were lying around and went sprawling. One of the Guardia men grabbed her hand and dragged her clear a fraction of a second before the 80-foot giant came crashing down.

It was now that George put through his SOS to TAC HQ in Panama City and I jotted down the alarming details on that pink slip of paper that happened to be on the desk in front of me. Immediately fixed-wing spotter planes and helicopters were scrambled, including choppers from the US Air Base in the Canal Zone whose previous emergency had been the mass suicide of the American religious sect in Guyana. Although George couldn't be certain of his exact location, he had worked out an estimated position which, as it turned out, was remarkably accurate. Even so, we knew that even if he was spot on it would still be quite tricky pinpointing him.

He was told to expect the planes to start arriving around 4.50 pm and as that time approached he called on his team for one final effort in keeping a bonfire well-stocked with greenery so as to create as much smoke as possible. This they did, but as the appointed hour came and went without straining ears being able to detect even the vaguest hint of an aircraft engine above the chatter of the parrots and parakeets, numbed minds were no longer able to spur aching limbs into action and everybody finally flopped. Keeping that fire alight was quite beyond them even though they knew it could be vitally important. An attempt to launch a marker failed because there was no water to spare to pour on the chemical which then reacted to produce the gas to fill the balloon. Urine, it was soon discovered, did not work! As a last resort Guardia private Luis Murillo was despatched to shin nearly 100 feet up the highest tree in the vicinity where he perched clutching a brace of red smoke flares which he was ordered to fire off if he saw anything flying his way.

Just as hope was fading along with the light and the dejected team roused themselves and set about rigging their hammocks for the night, Luis let out a triumphant yell and ignited the first of his flares as a large American helicopter tracked across the sky about a mile away to the South. The plume of red smoke was spotted by the crew and within seconds they were hovering 180 feet above the little clearing where the patrol were jumping up and down, waving and cheering with delight and relief.

The chopper then flew off and circled at about 2,000 feet whilst trying to get a fix on the position and there was another moment of panic on the ground when it lost them again – further evidence of the needle-in-a-haystack problems involved in picking out a clearing a few yards square amid hundreds of acres of identical jungle. An added frustration for those on the ground was that although they could see the plane above they couldn't easily talk it in on the radio because the link-up involved several relays – from the jungle to Caledonia Bay and then on to TAC HQ in Panama City, from there to the control tower at the Air Base by telephone and from the control tower to the plane – and the consequent delay in getting instructions through rendered them virtually useless.

After ten agonising minutes the helicopter re-established visual contact and returned to hover again at about 180–200 feet. The pilot was reluctant to land among the tree stumps and decided instead to winch Rik aboard. The winch was shaped rather like an anchor and the dazed and near-delirious cameraman was strapped to the flukes before being hoisted up. It was just as well that he was too far gone to appreciate exactly what was happening to him otherwise he quite likely would have died of fright!

Pausing only to promise that further flights would come out the next day to collect the remaining sick men and to bring out fresh water supplies the helicopter headed back to base. An offer was made to fly in another chopper that night and land it by torchlight but George would have none of that. With Rik safely away things were no longer quite so desperate, he said, and there was no point taking any further risks. The patrol settled down for a very thirsty night.

Amazingly, the first helicopter out next morning failed totally to locate them. It circled round and round the general area but was unable to pick out the smoke of their warning bonfire among other forest fires despite being less than two miles away at times and eventually it gave up and returned to base. On the ground dark despair set in again. Once more they had had to endure the frustration of being able to see the plane clearly without being able to attract its attention. What was worse, their frantic efforts to build up the fire in order to make more smoke had not only further exhausted them but had also caused them to sweat out extra precious liquid. They cut down vines in the hope of adding to the dregs of foul-tasting water that was all they now had left, but even these were all dried up and all they got were a few drops of moisture which did no more than dampen cracked lips. George was seriously thinking of breaking out the full bottle he had kept hidden at the bottom of his pack as a final emergency measure when, after four hours, one of the aircraft thankfully homed in on one of the Guardia's treetop smoke flares. But even then the wait wasn't over. The helicopter had almost run out of fuel and only had time to whistle up a fixed wing plane to mark the spot while it raced back to base to refuel and collect re-inforcements. This time the circling marker plane didn't let the clearing out of its sight. And within forty-five minutes a succession of choppers winched up the other sick men – they were whisked off to join Rik in Gorgas Hospital where all four soon recovered – and dropped in six jerry cans of water. The crisis was over.

The relief that flooded through them all once the crisis had passed is reflected in Desmond Dugan's diary entry for day 18, made after he and the rest of the team had collapsed into their hammocks exhausted but well watered and well fed at last.

'The events of the last few days will give me, and probably most of us, a whole new slant on life. We should now appreciate all those things we have taken for granted because they were at our fingertips. I will think back to today and remember it well every time I turn on a tap. Also, we should not forget how much effort has been put into the whole evacuation programme and the water drop. If it is not already expressed then I put it in the records now: from the depths of our hearts we thank all involved in coming to our aid. It was not a life or death situation but it so easily could have been.'

But it still wasn't quite over. There was another ten miles or so to go to Santa Fe and as the route lay through dense jungle it promised to be hard going once more. Furthermore, the level of water in the six jerry cans had dropped alarmingly in the few hours since it was dropped as people slaked their thirst and so a rationing system had to be reintroduced.

The next day was spent hacking through the undergrowth until about midday when the patrol hit upon a dry river bed that followed their route and progress speeded up. Even so, they had still not reached their goal by the time they camped that night, despite having been on the go for twelve solid hours.

Along the way they replenished their water supplies from two pools the contents of which one normally would have hesitated to use even for car-washing purposes. The first had dead fish floating belly-up on the surface while the second had obviously been wallowed in by a herd of peccari – a type of wild pig – and stank to high heaven. However, with the experiences of the previous few days very fresh in everybody's mind the murky liquid which they happily proceeded to filter through their Millbank bags and then boil up somehow seemed as welcome as a crystal clear mountain stream.

The evidence of a herd of peccari was interesting. Rather like piranha fish, these small pigs are pretty harmless individually but can be deadly in a herd. Locals take to the trees if they hear them coming through the jungle en masse and they say that if they besiege you in the branches your only means of escape is to throw down an article of clothing in the hope that they will eat that, assume it is you and move on. People who have been caught in the path of a fast-travelling herd of these vicious creatures say it is a terrifying experience.

Luckily the herd that had visited the jungle pool was no longer in the vicinity, so that was one danger the patrol didn't have to face. There was a truly relaxed atmosphere in the camp that night for the first time for many days as people contemplated the pleasing prospect of a leisurely walk down to

Santa Fe the next morning to bring their mission to a successful end. After a meal of turkey and rice they chatted round the camp fire and then retired to their hammocks where they gazed up at the stars and reflected on what they had been through.

For those who had made it there was a great feeling of achievement. Suddenly all the hardships were something to remember with pride and there was satisfaction in the knowledge that dangers had been faced and overcome.

The experience had given some interesting insights into human nature under stress. For instance, it was seen that those who were physically the strongest didn't necessarily have the greatest mental resilience and therefore cracked up more easily. One thing which stood out a mile, and which was noticed by everyone, was the way the three girls came through it all. No concessions were asked or given and they not only matched the men stride for stride but showed a calmness under pressure that proved an invaluable steadying influence for everyone else. Cathy Davies, a Scottish international basketball player who was just seventeen at the time, proved a tower of physical, as well as mental, strength. When some of her male colleagues were wingeing at the thought of having to hump a full jerry can of water she just stepped forward without a word and hoisted one onto her shoulder as if it weighed nothing. Despite being one of the youngest members of the team she was put in command of one of the three groups into which the YE's were split and emerged as a natural leader. Canadian student Ann Smith and English secretary Denise Wilson also made deep impressions on everybody through their strength of character, their determination to see it through on equal terms and, not least, their feminine charm. It was generally admitted by the men that it was the three girls who held them together at times.

On 16 March – day 20 – the patrol reached the Pan American Highway – a ridiculously grand-sounding name for what was at that point little more than a 25 foot-wide dirt track – and prepared for the triumphant march into Santa Fe. This was the occasion for the final confrontation between Juan Diaz

– whose immediate thought was to hitch a lift for the last couple of miles – and George Thurstan, who was adamant that his men had started out on foot and were going to finish that way. Just when it looked as though the rivalry that had been simmering between the two of them throughout the trek was finally going to boil over, Alan Bibby and his film crew drove up in a truck and dispensed lukewarm coca cola and beer and the bubble of tension burst.

George relaxed and gave the order: 'Form a rabble!' whereupon the three teams of YE's actually raced each other into Santa Fe. Needless to say Cathy Davies won!

The next day they either flew or sailed down to La Palma to await transport back to Caledonia Bay. Dreams of hot showers and clean clothes at La Palma were shattered when they arrived to find a water shortage. As a consolation they were loaded into a truck and taken some miles to a quarry where a couple of pipes had been rigged up so as to provide a pathetic dribble of water from a natural spring. Hardly the thing to wash away the ingrained grime of three weeks on the trail. But somehow it seemed a fitting end to an exploit in which physical discomfort and hardship had been a key factor as those taking part drove themselves into the ground to such an extent that most of them did little but sleep for a fortnight afterwards, whilst George Thurstan admits that it took him six months to recover fully. But that, as he says with a certain masochistic pleasure, is what it was all about.

Chapter 6

Eye of the Wind watch leader Chris Sainsbury and his three YE companions stared in stunned disbelief at the dusky Polynesian beauty sitting demurely in their midst – but there could be no doubt about what she had said; she would like to go to bed with them – all of them. Chris is not a man often to be lost for words, but for once his mouth was opening and closing like that of a feeding goldfish as he searched his mind for a suitably diplomatic refusal to her matter-of-fact request.

It had all started innocently enough. The *Eye of the Wind* had sailed out of a spectacular South Sea sunset and into Tahiti's Papeete harbour in fine romantic style, with her crew aloft and an escort of ten-man canoes, and after savouring these moments Chris had led his party ashore for a look around the town. It was while they were inspecting the open-sided restaurant vans that lined the quayside that they fell into conversation with the girl who introduced herself as Stella and explained that she was the daughter of a local politician. She showed great interest in the *Eye of the Wind* and proved a fount of knowledge about local history and customs as well as a source of fascinating gossip about some of the other ships in port.

In the end she insisted on buying everybody dinner and, as the four men gorged themselves on sticks of roasted cow's heart, they wondered at her generosity. What a marvellous gesture of friendship. How could they repay her? Even as they asked themselves the question they got the unexpected answer – the simple matter of a favour ... or four!

After groping for the right phrase with which to decline this proposition without causing offence, Chris pointed out, as tactfully as possible, that unfortunately they were broke and therefore couldn't take advantage of the offer, whereupon Stella let out a shrill cry of protest and abuse. She never did 'it' for money, she insisted, but only because she enjoyed it! Even allowing for the fabled sexual hospitality of the South Sea Islanders, Chris and his boggle-eyed YE companions thought this was unlikely and beat a hasty and somewhat undignified retreat to the ship.

Their encounter was just one of the many bizarre and exotic adventures that helped to make the voyage of the *Eye of the Wind* across the Pacific from Panama to Fiji one of the most colourful of Operation Drake's nine phases. A real Treasure Island was visited, as well as a remote atoll whose inhabitants all share the same name, having descended from one 19th century British adventurer. Weird and wonderful creatures were encountered and in between times life on board was often – though not always – idyllic.

It began on 21 April when the ship – with most of her expensive mechanical troubles at last thankfully behind her – slipped her moorings in Panama and sailed through the canal bound for Costa Rica.

For the YE's the first thrill of the trip came when acting skipper 'Tiger' Timbs hove to and allowed them to plunge into the inviting Pacific water for a swim. Some of them enjoyed it so much that luring them back aboard proved quite a problem. One who got quite carried away was Tyrone Spence from England. He had only ever swum in the sea once before in his life and that was during a Butlins holiday at Skegness. Not quite the same as diving from the deck of a sailing ship into the warm Pacific! He came out spluttering his excitement. 'It's so deep, so blue, so clear – wow! What an experience,' was all he could gasp when reprimanded for ignoring the whistle that summoned everybody back on board.

The next day a party of eighteen YE's under the command of watch leader Frank Esson and marine biologist Patricia Holdway were dropped off at San Pedrillo for the start of a four-day foot patrol through Costa Rica's Corcovado National

Park while the ship went on ahead to Drake's Bay on the Osa Peninsula where two special plaques were to be unveiled. The plaques – presented by the Costa Rican government and the Lord Mayor of Plymouth – commemorate Sir Francis Drake's visit to the Bay in 1579. The *Golden Hind* was at that time already heavily laden with treasure captured from the Spanish galleon *Cacafuego* but, while she sought refuge among the sheltered islands and hidden bays of the area and prepared to return to England across the Atlantic, Drake sallied forth to take another prize of a very different nature. Aboard the Spanish cargo vessel which now fell into his clutches were two pilots armed with charts of the 'China Run' established by Spain across the Pacific between Mexico and the Philippines. This was an even more valuable windfall in its way than the *Cacafuego* gold since it was the key to the new way home via a Southern route that Sir Francis had dreamed of. Pausing only to careen the ship and establish another Drake's Bay near the site of what was to become San Francisco – where he claimed what is now California for the Queen and named it New Albion because the white cliffs of the coast reminded him of Dover – he set sail across the Pacific.

Torrential rain delayed the planes that were flying in some of the VIP's for the unveiling ceremony, but it all went well in the end as the British Charge d'Affaires and the Costa Rican Director of Tourism shared the formalities. Owing to a shortage of local skilled labour, the Ministry of Overseas Development representative, Mr Gordon Rayson, had built the memorial into which the plaques were set with his own fair hand – and a very good job he had made of it. After the ceremony the sixty or so guests lunched aboard the *Eye of the Wind.* The white-painted ship looked particularly magnificent as she rode at anchor, framed against the startling blue of the sea and the vivid green of the jungle.

She remained there for nearly a week while the YE's took part in various brief land expeditions aimed at giving them a first taste of the tropical rain forest – not to mention its many exotic fruits including pineapples, bananas, sugar cane, sweet lemons, mangoes, coconuts, limes, avocados and breadfruit.

From there it was on to the little island of Cano – or Caines, as it was once known – another of Sir Francis's hiding places whilst he was in that part of the world. It was fascinating to reflect, as one stood on deck and gazed out across the little bay at the sandy beach with the tree-clad cliffs rising sharply behind it, that this was almost certainly the view that Drake himself looked out upon 400 years before. There was no other suitable anchorage he could have picked as a sanctuary while his men careened the ship and took on fresh supplies of food and water. These days the island is uninhabited except by the ghosts of the many indians who are buried there. The mainland tribes used it as a burial ground some 800 years ago and although hundreds of the graves have been robbed by treasure hunters hoping to find gold and jewellery interred with the bodies, many more still remain untouched. The island's other great attraction is that its climate is apparently ideal for the growing of marijuana! This affords a marvellous opportunity for those who don't want to risk being caught with a crop of the illicit weed in their own back yard to sneak ashore after sailing from the mainland, sow the seeds and return later to reap the harvest. Maybe that was why a Costa Rican naval gunboat took such a close interest in the *Eye of the Wind*. A boarding party was actually sent over but was recalled even before it reached the ship, presumably because some more pressing emergency had arisen. This was probably just as well since 'Tiger' might have experienced some difficulty in explaining how he and his motley crew happened to be there without the proper immigration papers or cruising permit. Operation Drake's presence in the area was covered only by the vaguest form of diplomatic immunity and the fact that the Costa Rican equivalent of the Admiral of the Fleet had attended the unveiling ceremony at Drake's Bay would not necessarily have cut a lot of ice with suspicious security forces who very likely didn't speak a word of English.

From Cano the *Eye of the Wind* headed back to Panama to pick up her new captain, Mike Kichenside, before embarking on the 8,000-mile voyage across the Pacific. On the way she made an ocean rendezvous with the BP supertanker *British*

Resolution – ten BP cadets had joined us during Phase Two and had used the *Eye of the Wind* as a training base – and the contrast between the two vessels was startling. Suddenly the *Eye of the Wind* was reduced to the scale of a child's Serpentine model as she sailed around the giant tanker.

Mike Kichenside's arrival on board to take over formally from 'Tiger' – who had greatly enjoyed the brief interlude as skipper of his own ship – was eagerly awaited. But while everyone on the ship was immediately impressed by him, he must have wondered what he had let himself in for as he surveyed the sorry-looking bunch that paraded before him. They resembled a walking casualty department with cuts and bruises and slings and bandages everywhere. The reason for this was a rugby match that had taken place the previous day following a challenge from the local army club. The 'game' had resulted in more injuries than had been sustained in the whole of Operation Drake up until then.

The trouble was that among the *Eye of the Wind*'s international crew of YE's only three knew the rules of English rugby while the rest operated on the simple basis that if a member of the opposition got the ball he should be eliminated as swiftly as possible, using whatever technique came naturally! With Australians, Americans, Canadians and Tasmanians among the Operation Drake XV, the other side were ruthlessly felled with tackles that combined the roughest features of rugby league, American football and ice hockey. It wasn't long before they resorted to similar tactics and the pitch soon resembled a battlefield with the wounded strewn out from end to end. Operation Drake limped away from the battle claiming victory while their opponents suggested that maybe the local SAS team would be a better match for them!

The first six days of Mike's command were fairly uneventful as the *Eye of the Wind* crawled sluggishly towards her first port of call, at Cocos Island, in the lightest of breezes. Despite the calm there was a heavy swell and sickness was again rife as the YE's struggled to get their sea legs. Apart from that, the only upheavals came from the decision of the YE Council to hold its committee meetings in camera. This greatly upset

Capital Radio's Pam Armstrong who had been hoping to tape the proceedings for broadcast as an example of democracy in action! For a short time everything got very political but once the pace of things hotted up there were plenty of more interesting distractions to occupy people's minds.

Cocos Island got its name because of the coconut groves which the early explorers found already flourishing there when they first arrived. As the place was uninhabited the question of who planted them has always been something of a mystery – although Thor Heyerdahl suggested that maybe Polynesian navigators were responsible. His theory was that they needed a source of unripe coconuts which they could call in for and take away as a source of water on long voyages. The groves have long since vanished, but the name remains.

Cocos has a colourful history that has led to its reputation as a real life Treasure Island. The pirates Henry Morgan and Benito Bonito – otherwise known as Bennett Graham – are said to have used it as a hiding place for spoils worth millions of dollars that were looted from ships and cities along the South American coast. Even more precise is the story of the so-called 'Treasure of Lima' – entrusted to Captain William Thompson of the *Mary Dear* by the authorities of the city during a revolution in 1820. Thomson is said to have buried the gold and jewels somewhere in the vicinity of the island's Wafer Bay. Estimates of the total value of the hidden booty range up to $100,000,000 and have attracted endless bounty-hunting expeditions over the years.

It wasn't too hard to conjure up visions of a buccaneering past as the *Eye of the Wind* slipped into Chatham Bay where the water was so crystal clear that Captain Kichenside was able to pick a sandy spot amidst the rock and coral upon which to drop anchor. There is an air of fascination and romance hanging over the place which is undoubtedly why so many European beachcombers, artists and lets-get-away-from-it-all drop outs are attracted there.

Not everything about Cocos was entirely charming as 'Spider' Anderson and 'Tiger' discovered when they dived overboard to inspect the propeller only to find themselves under close scrutiny from five sharks – two of which they estimated to be nearly 15 feet in length. Chris Sainsbury, who had never seen a shark before, later dangled a frozen fish over the side as bait and was swiftly rewarded when there was a dark blur in the water which left the hook dangling bare.

During the 36-hour stopover at the island the YE's divided into three parties. One, under Frank Esson, went in search of the wreckage of an American fighter plane which had crashed into the island's highest mountain and then lain undiscovered for thirty years on the densely-forested slopes. They succeeded in finding it, but, judging by the shattered looks on their faces when they got back after the rushed climb, most of them probably wished they had heeded Frank's light-hearted warning that his group were in for 'utter misery and total deprivation' and had chosen instead one of the softer options. These involved either going with Patricia Holdway in search of a particularly rare type of stick insect and other specimens of local wildlife or helping to make a detailed survey of the flora and fauna of the shoreline. Here the greatest fascination was provided by the nesting Blue-Footed Boobies. These birds are so tame that one can wander among them at will without provoking any reaction other than an occasional warning hiss. It was as well to beware of the hoards of marauding land crabs that surrounded every nest waiting for an opportunity to steal either the one egg, or the young chick, which each pair of birds produce.

'Spider' added the name of the *Eye of the Wind* to all the others that have been carved on the rocks of Chatham Bay – the traditonal visitors' book of the island since 1757 when a passing mariner first obeyed the age-old urge to leave his mark – and then the ship set off on the next stage of her voyage. Just before she left a torrential downpour had every member of the crew dashing up on deck with every available receptacle from saucepans to dustbins in an effort to catch every possible drop of fresh water and replenish supplies the simple way.

The four-day haul from Cocos to the Galapagos was made sluggish by a constant swell and heavy current that slowed progress to a frustrating four knots, but the daily routine on board was enlivened by preparations for a grand crossing-the-line ceremony. A strange silence settled over the ship as almost the entire crew retreated into corners and, brows knit with concentration, chewed their pencils to the lead as they waited for the muse to settle and give them the inspiration for yet another lewd limerick to accompany each individual 'initiation'. 'Spider' spent hours working on props and costumes, 'Tiger' would every now and again start cackling with glee and rubbing his hands together with delight for no apparent reason as he dreamed up another suitable 'punishment' and in her laboratory Patricia Holdway lovingly prepared some truly revolting mixtures to be poured over the thirty initiates who would be crossing the equator for the first time.

It was on 21 May that King Neptune's court assembled on deck and the proceedings commenced. YE Gillian Rice took the part of Neptune while Chief Petty Officer Richard Shrimpton – the ship's engineer whose Royal Navy blue verses had to be heavily censored – made a startling appearance as Queen Amphitritos. Watchkeeper Robert Clinton was Clerk to the Court – very apt since he is a lawyer by profession – and YE Bruce Buckley was appointed Court Barber. 'Tiger' and purser/nurse Leslie Reiter were Court Attendants and the role of Davey Jones was taken by Mike Kichenside's twelve-year-old daughter Miranda who has sailed everywhere with her father ever since her mother was tragically killed in a car accident.

The ceremony went off in the traditionally boisterous manner complete with rude rhymes, duckings and the administration of some disgustingly messy punishments. How Chris Sainsbury managed to survive the nauseating experience of being force-fed through a plastic tube with large quantities of gin mixed with cooking oil remains one of the great mysteries of Operation Drake. In the end nobody escaped unscathed and even the skipper submitted gracefully to a ducking. The day ended on another note of excitement when the *Eye of the Wind* anchored that night within sight of the twinkling lights of Wreck Bay, Chatham Island – the first port of call in the Galapagos.

The archipelago is composed of sixty-one small islands and straddles the equator some 600 miles off the coast of Ecuador. It was here in 1835 that Charles Darwin – then twenty-two and therefore about the same age as our Young Explorers – made the observations which formed the basis of his revolutionary *Origin of the Species*. The YE's were as fascinated as Darwin had been by the wide variety of flora and fauna and especially so by the unique ways in which some of the species have adapted to their environment.

In order to make the most of the ten-day stopover, Mike Kichenside sailed between the various islands during the night and left the daylight hours totally free for the many land expeditions that were mounted. These ranged from pure sight-seeing trips to more serious scientific projects involving census-taking and survey work on behalf of the Charles Darwin Research Institute.

Among the highlights of the visit the one which seemed to catch the imagination of the YE's more than any other was the unique experience of swimming with the sealions off South Plaza island. Like dolphins, these creatures are playful, inquisitive and apparently quite at home in human company and it wasn't long before they were cavorting happily in the water with the YE's, most of whom were delighted by the novelty. It more than made up for the disappointment of not being allowed to ride the giant tortoises they saw at the Charles Darwin Research Station on Santa Cruz. Many of them had hoped to hitch a lift astride one of these huge, lumbering beasts just as Darwin himself had done, but, although touching and feeding was permitted, joyriding was ruled out of order by Station officials.

The barren volcanic island of Marchena was memorable for the extraordinary lava flow that looked for all the world like a swollen, rushing river that had suddenly been turned to stone at the wave of a magician's wand. Standing amid the rather eerie landscape of this lonely island – one of the few that is rarely visited by the tourists who flock everywhere else – the haunting mystery of the so-called Countess of the Galapagos no longer seemed so unlikely.

The notorious Countess arrived to live on nearby Floreana island with two young men, one of whom, an Austrian, disappeared only for his desiccated body to be found later on Marchena along with that of a fisherman. When the authorities went to question the Countess, she and her other boyfriend had also vanished. Neither was ever seen again – at least, not officially. The story – fully recounted in the diaries of a lady called Mrs Wittmeyer who still lives on Floreana – has a spine-tingling tail piece. A few years ago – a long time after the original drama – a charter boat made one of its regular visits to the island and among the tourists who came ashore to look around was an old lady. But when the time came for the boat to leave she was nowhere to be seen. The Ecuadorians searched the island for three weeks using helicopters and troops but no trace of her was ever found. She had disappeared into thin air. It was then that one or two people recalled that she did have a very aristocratic bearing and began to wonder whether maybe it was the Countess making a last macabre pilgrimage to the scene of the tragedy.

Operation Drake was able to make one very positive and valuable contribution to wildlife conservation on Galapagos. On Santa Cruz the land iguana population has been sadly decimated over the years by the ravages of wild dogs, to the point where these weird and wonderful reptiles are in danger of being wiped out. The Darwin Research Station had already launched a rescue programme whereby breeding grounds on the island would be wired off to keep the dogs out, but they were having difficulty in finding any females with which to stock the sanctuary. Exhaustive searches had failed to unearth the elusive ladies and the situation was getting rather desperate by the time the YE's arrived on the scene.

They immediately set off to investigate the largely deserted burrows and after much peering down dark, empty holes Gillian Rice found her gaze being returned by a pair of beady yellow eyes! With the help of her colleagues Mark Greer and Cathy Lawrence from Canada and Simon Turton from England she started the task of digging the animal out. This proved easier said than done as the iguana retreated back into the burrow as fast as they could tear away the turf and earth with their bare hands. But half-an-hour and five foot of excavation later their quarry began to emerge and, pausing only to bare its fearsome-looking teeth and hiss aggressively, made a bolt for it. Before it could make the safety of some nearby bushes it was bravely pounced upon and trapped underneath a jacket. It was then gingerly transferred to a haversack and borne triumphantly back to the Station where it was confirmed as a she – and a very fine three foot specimen at that.

There was just one other duty to perform before the *Eye of the Wind* finally left the Galapagos and sailed off on the next leg of her voyage, and that was to go to Post Office Bay on Floreana and pick up the mail in the time-honoured local tradition.

There is a gaily-decorated barrel on the quayside which passing ships have used as a post box for nearly 200 years. The way the system works is that each vessel deposits her own mail and picks up whatever is addressed to any of the places she is headed for. The *Eye of the Wind* duly left behind forty letters and cards and collected everything that was addressed to the South Sea Islands. And then she headed out into the wide blue yonder of the Pacific.

It was almost exactly a month before those aboard set foot on dry land again, even though the winds were strong and favourable and pushed them along at a cracking pace that averaged around six knots. Those who had anticipated sun-bathing conditions were soon disillusioned – sweaters, and even oilskins, were the order of the day as fresh breezes whipped the spray across the decks. But it was very invigorating and now that seasickness was no longer a problem nobody was complaining too much – nobody except poor Patricia Holdway. She tried every day for a week to give a lecture, with slides, on the plight of the whale but was forced to give up each time because there was no way of keeping the projector and screen on the level as the ship lurched through the water. In the end she was able to give her talk with 'live' illustration when several schools of sperm whales passed nearby.

Patricia also found the fast pace frustrating because it resulted in her sample nets being ripped to shreds as they were trailed behind the stern. Nevertheless, she did manage to trawl a number of interesting specimens including a rare nautilus. These snail-like creatures live in the deepest parts of the world's oceans and the most one ever sees of them normally are their shells, which float to the surface when they die and are worth a great deal of money. To find a live one was quite a coup for our marine biologist.

For everybody else life on board soon settled into a fairly smooth and easy routine. When not actually on watch duty, the YE's were kept busy with lectures, of which those on navigation proved immensely popular. Socially, the highlight of each week was the special Saturday night dinner party laid on by successive watches. The first of these was a French Evening, when amazing gastronomic delights were conjured up in the galley, and this was followed by a Trappist Monks Evening which turned out to be highly entertaining as everybody set about enjoying themselves without actually speaking a word. The night ended on a hilarious note with a joke session conducted entirely with cue cards.

The first sighting of land came as the ship approached the Tuamotu Archipelago and there was considerable nervous excitement among the YE's as a result of the terrifying yarns 'Tiger' had been spinning about the many ships that had been wrecked among the treacherous atolls which protruded just a few feet above the water like jagged teeth. But if the atolls were menacing, the islands themselves were breathtakingly beautiful. With their golden beaches fringed with palms they lived up to everyone's expectations of what a South Sea island should look like.

After the brief stop in Tahiti – during which the alarming state of the propeller was discovered and repaired – the *Eye of the Wind* pressed on to Fiji. On this last lap of the voyage the fair winds finally deserted and the ship found herself frequently and frustratingly becalmed, with the result that visits to Tonga and Palmerston Atoll had to be cancelled in order to make up lost time. This was a great pity since the King of Tonga had been planning to go aboard the ship – and Palmerston Atoll has a fascinating history.

Originally discovered by Captain Cook in 1774, it was uninhabited until the middle of the nineteenth century when one William Masters settled there with his three Polynesian wives. He had left England to join the California Gold Rush of 1849 and it was after much globe-trotting that he ended up on the tiny atoll – seven miles long and five miles wide – and decided to make it his home. Before he eventually died in 1889 he had founded a considerable family and his son carried on the good work.

By the time an American round-the-world sailor called in in the 1920's the little colony was over 100 strong. Masters' son, having married several times, was at that stage in the interesting situation of having a wife who was younger than his grand-daughter! Some of the relationships within the community were mind-bogglingly complicated – but, strangely, the in-breeding had not caused any obvious degeneration. Another fascinating feature of the little society was its language – an antiquated form of English that was a century out of date.

A later visitor to the atoll wrote about a freak hurricane in 1926 followed by a tidal wave during which the entire community took to the trees. While they were up there in the branches one of the women gave birth to a daughter!

In view of this colourful past the YE's were very disappointed that they were unable to visit the place, but there was a consolation. Some of the Islanders came out to the ship in their canoes as she passed by and exchanged presents. Among them was the Island's radio officer who introduced himself as ... John Masters.

The *Eye of the Wind* reached Fiji on 18 July – eighty-nine days after she left Panama. For all those aboard it had been an epic voyage and, perhaps more than any other phase of Operation Drake, it welded everybody involved together in a spirit of comradeship that developed naturally from living together in such confined conditions for such a long time.

The changeover of the Phase III and Phase IV YE's took place in Suva and the new arrivals went straight to Moala – an island ninety miles south east of Fiji – where they spent a fortnight giving invaluable assistance to the local hurricane relief project. Hurricane Meli had struck the Lau group of islands with devastating force just four months earlier and had left a trail of destruction. Forty-nine people were killed and 1,200 left homeless as winds of over 100 mph ripped buildings from their foundations and hurled them through the air like matchboxes. Crops were flattened and the local fishing industry brought to a standstill by the destruction of all the boats.

Moala was by far the worst-affected of the islands and it was here that Operation Drake concentrated its efforts. Even in the short time that was available the twenty-eight YE's and directing staff were able to erect school classroom blocks and teachers' quarters in the villages of Keteira and Vunuku. It was hard work, but they were rewarded by being treated as honoured guests by the grateful villagers who had been unable to get any real help before Operation Drake arrived on the scene.

While all this was going on the *Eye of the Wind* was back in dry dock – this time to have the prop shaft replaced. Thanks to the magnificent combined efforts of the Royal Fiji Navy, the local shipyard, Government officers and local supporters the job was done as quickly and efficiently as possible. But, once again, the extra expense had our accountant shaking his head in despair.

At this point we had one of our rare strokes of good fortune as far as the ship was concerned. Columbia Pictures were in Fiji filming a remake of *The Blue Lagoon* and they asked if they could hire the *Eye of the Wind* for a few days. Not only did this provide an interesting diversion for the crew – it also helped quite considerably towards the cost of the repairs.

On 16 August the ship set sail at last for Papua New Guinea – the second largest island in the world after Greenland and one of the least-explored places on the face of the Earth. Not for nothing has this inhospitable jungle-clad land of mystery been dubbed 'The Last Unknown'.

Chapter 7

The giant Chinook helicopter lifted off and rose above the tiny jungle clearing with a deafening roar from its powerful engines, while the downdraught from its twin rotors created a mini-hurricane that raised a swirling cloud of dust and whipped the surrounding foliage into a frenzy of dancing branches and wildly fluttering leaves. On the ground fifteen near-naked brown bodies prostrated themselves, their hands clamped over their ears, their foreheads pressed to the earth, their eyes not daring to look up. As the scene receded out of focus below him, Major Roger Chapman peered down anxiously at the prone forms. They were primitive people who had used stone axes to help him and his team clear the landing site for the chopper without having the faintest idea of what was about to descend on them. What must be going through their minds now, he wondered.

Up until just a few hours before they had never set eyes on a white man and had had no contact with modern civilisation. Then suddenly they had been brought face-to-face with the 20th century in the most dramatic fashion. No sooner had this party of strange-looking and oddly-dressed men arrived in their village, in an inaccessible spot in the remote Central Highlands of Papua New Guinea, than they had started speaking into some sort of magical box and had summoned up the amazing mechanical monster which then whisked them away into the skies. Fifty years ago – when aircraft first arrived in Papua – similar encounters had given rise to the Cargo Cults. It is fascinating to speculate about the legends that may grow up around the meeting between Chief Kemba's little tribe of nomadic Pogaians and Roger Chapman's Operation Drake recce party. The event will undoubtedly pass into local folklore as the tribal storytellers raise their sing-song voices over countless camp fires and spread the news in the traditional way to the remotest corners of the forest.

It all happened when Roger led an advance party on a gruelling twenty-six-day patrol through the largely uncharted Strickland Gorge in preparation for a first-ever attempt at tackling the white water challenge of the Upper Strickland River. Only a handful of white men had ever seen the seventy-five mile stretch of rapids and cataracts that comes cascading through the 2,500 foot cleft in the mountains and nobody had thought seriously about going down it in a boat. Roger's plan was to take a party of eight men through in two inflatables. Bur first he needed to go and have a close look to make sure that the venture was at least vaguely possible. The aerial survey he had already made was no good by itself because you cannot really judge a rapid by the two-dimensional view you get from above. Apart from that, large stretches of the river are permanently hidden under low cloud so that maps of the area – which are compiled from aerial photographs – contain white patches bearing the legend: Cloud – Relief Data Incomplete. That's the modern equivalent of the mythical creatures, the gogs and magogs, with which the old cartographers used to fill the unknown areas they had been unable to get to. Papua New Guinea is one of the very few places on Earth where the maps still have blank spaces.

Roger's research showed that only four white patrols had ever ventured into these regions before – the first one only twenty-five years previously. All four were led by Australian Kiap Officers – the equivalent of the British Colonial District Commissioners – in the days before Papua was granted Independence in 1975.

The real pioneer was Des Clancy who, in 1954, went in with three representatives of the Australian Petroleum Company and 150 bearers, guides and armed escorts to search for oil shales near the Strickland Valley. They followed the course of the river as closely as possible along the Gorge and eventually carried on through to the lower reaches where it broadens and meanders more sedately for some 430 miles before joining up with the Fly River. They never did strike oil – but Clancy came back with some interesting information about what lay hidden under the blankets of cloud that obscured parts of the Upper Strickland. At one point – a location called the Kagwasa Crossing – the whole of the mighty river is channelled through a deep split in the granite no more than 10 feet wide. It was, he reported, an awe-inspiring sight. Later, he was forced to climb out of the Gorge because of the terrain, but, from a 3,000 foot vantage point on the high ground, he got a distant view of the gap where the water gushed out of the mountain ranges into the plain below. The mist and cloud that rose like steam above it caused him to christen the spot 'The Gates of Hell' while the churning white water rapids beyond he named 'The Devil's Race'.

Two years later another experienced Kiap officer, J. P. Sinclair, set out with one white companion and 106 bearers intending to cross the river at the Kagwasa Crossing. But when he arrived there, he was greeted by a sight even more astonishing than that which had taken Clancy's breath away. It was the rainy season and the river had risen more than 20 feet so that the level was actually above the narrow fissure – but the line of it was marked by a fantastic mare's tail of spray hurled many feet into the air by the forces created under the surface. So great were these forces that hefty logs were being thrown up and then bobbed like table tennis balls on the crest of the fountaining jets of water.

The two other patrols that ventured into the area between 1968–72 struck out into the interior of the Muller Mountains with the specific aim of trying to contact the native inhabitants and count them for the national census. Roger actually managed to track down members of each of these teams. They were able to give him very useful briefings about both the conditions and the people he could expect to encounter.

They confirmed that the terrain was as hard and inhospitable as one was likely to find anywhere in the world.

When you weren't scrabbling up and down almost sheer-sided ridges, while at the same time having to hack a passage through dense undergrowth, you were likely to be picking a path across limestone rock so sharp-edged that the toughest boots would be shredded after a few miles. Even the seemingly open areas that appeared from the air like smooth carpets of green were actually covered in 15 foot high Cunhai grass that was so difficult to cut through that a fit man would be reduced to sweat-soaked exhaustion within a matter of yards.

As for the natives – many of them were extremely primitive. The extent to which the tribes were cut off from each other – let alone the outside world – was reflected in the fact that at the last count there were at least 700 known dialects throughout the island. So impenetrable was the jungle that communities in neighbouring valleys were often found to be unaware of each other's existence. Their reaction to white men was unpredictable. The early patrols had adopted a policy of sending ahead shouters – native guides who would go forward and literally bellow out advance notice of the party's arrival in the area and call in any tribes in the locality for a meeting. Salt and tobacco would then be handed over as a gesture of friendship and, once the ice had been broken, impressive shooting displays with rifles would be set up to discourage any hostile notions. There had never been any attacks although one of the census patrols had narrowly avoided an ambush.

It was against this background that Roger led out his ten-strong recce party. Apart from himself this included ex-SAS signaller Sergeant Dave Weaver, an Australian expatriate business man, Bob Woods, and American Bill Neumeister, a very tough character who normally makes his living as a smoke jumper in Alaska – a job that involves being parachuted into the heart of forest fires to put them out before they really get started! There were also four native bearers and two members of the Papua New Guinea Defence Force who went along to act as liaison officers.

The bearers were recruited from villages around Lake Kopiago – base camp for the expedition – but finding suitable men was not easy. There was a marked reluctance to venture deep into the mountains, partly because of the punishing terrain and partly because of the risk of straying into enemy territory. However, with the help of the District Interpreter and a show of hard cash, resistance was finally overcome and four fine, strong-looking individuals were taken on.

They were all from either the Duna or the Huri tribes – otherwise known as the wigmen on account of the extraordinarily elaborate head-dresses they wear. To see the males strutting around town on market day like peacocks while their womenfolk get on with the business is quite an experience. It is like watching an exotic Easter Bonnet Parade. Each crowning glory seems more extravagant that the one before – towering structures of hair, feathers, fur and flowers that are often shaped rather like Napoleonic hats. Quills or bamboo are traditionally worn through the nose but, since the advent of Western influence, the in-crowd favour ball point pens!

Roger's bearers decided to leave their wigs at home for the trip and made do instead with a selection of tropical flowers stuck into their magnificent heads of thick, woolly hair – the natural afro-style mops that gave their island its name. The first Portuguese explorers took one look at the inhabitants and christened it 'Ihlas de Papuas' – 'Island of the Fuzzy Haired'.

It didn't take long to discover that the going was every bit as tough as everybody had warned it would be. Within three days – before they had even reached the river – the party lost the services of the two Defence Force men, who had to return to base when their legs gave out under the strain and badly-swollen knees threatened to bring them to a complete and painful standstill. Everybody else was suffering too as they sweated, slithered and clawed their way through the forest, stumbling under the weight of their 50 lb packs as their feet caught in the tangle of roots and creepers. Even the bearers found it hard to cope at times as the patrol pressed on at a murderous pace. They were on the march by 6.45 am every morning, in an effort to make the most of the cool of the day, and didn't call a final halt until 7.00 pm when they collapsed, exhausted, into their hammocks. By that time they would be so tired that neither the mosquitos nor the regular, drenching downpours disturbed their slumbers.

Once they had reached the river it was a matter of following its course as closely as possible, carefully charting every feature and grading each rapid from 1 to 7 depending on the degree of danger and difficulty it presented.

After three weeks the party had reached a point just to the North of the Bulago River junction. They had already passed the Kagwasa Crossing a few days before – and found it every bit as awesome as Clancy and Sinclair had promised – and were now about to enter one of the mysterious blanks on the map. Having camped for the night on a narrow ledge above the river, they awoke to find that the torrential rain which had bucketed down while they slept had transformed their surroundings. The level of the water had risen so dramatically that they were effectively cut off on their ledge and could neither advance nor go back the way they had come. There was only one way out – and that was upwards. It meant a difficult and dangerous 3,000 foot climb up the side of the gorge that was almost sheer in some parts – but there was no other alternative.

It took them nearly ten hours to scramble their way to the top. They had to move with the utmost caution, hauling themselves up on vines, roots and tenuous handholds in the crumbling rock. By the time they finally made it they barely had the energy to rig their hammocks and prepare a meal before collapsing for the night.

The next morning they decided to strike inland, away from the gorge, in search of a suitable clearing in which to take a re-supply drop from the air. After cutting through the jungle for about three hours they picked up an old hunting track and followed that for a further couple of hours until they reached a stream which provided a welcome excuse to stop for a cup of tea. It was as he bent to scoop a billycan of water for the brew that Roger noticed something he had not expected in such a remote area – the distinctive smell of wood smoke from someone else's campfire.

The patrol immediately crept further along the track until,

rounding a corner, they glimpsed through the trees a small clearing in which stood two low, rectangular huts with wisps of blue smoke filtering through the thatched roofs. Then figures were seen moving about and the sounds of children playing and pigs squealing floated across the forest.

Roger sent ahead two of his Duna bearers – Harrega and Aiyipee – as an advance party and when he saw that they had received a friendly welcome he led the rest of the patrol forward into the clearing. As he did so Aiyipee directed towards him one of the men he had been conversing with in a strange dialect and when Roger extended his hand the small, bearded man grasped it firmly and held on, obviously not wanting to let go.

There then followed a four-way conversation. Roger spoke in English to Bob Woods, Bob translated into Pidgin – the nearest thing to an official national language in Papua – so that Harrega could then rephrase the message in Duna for the benefit of Aiyipee who finally rendered it in the strange new dialect that was unfamiliar to everyone else but which was understood by the tribesmen.

In this rather tortuous, roundabout way it was learned that the village was called Tigaro and the inhabitants were nomadic Pogaians. There were five families in the group which numbered fifteen altogether. The man who was hanging on to Roger's hand so affectionately was called Kiwanga and the Chief – who was at that moment working in the village 'gardens' some way off but was expected back at sunset – was named Kemba. When he duly arrived he was able to confirm that nobody in his small tribe had ever seen a white man before – a fact which seemed to be amusingly borne out by the way in which they subjected Roger, Dave, Bill and Bob to a close scrutiny and the occasional curious prod.

The fascination was mutual. Kemba and his people were clearly untouched by modern civilisation. Their most sophisticated – and prized – possessions were two steel axe heads which they had traded with other tribes in the Bulago Valley where they had been introduced originally by one of the later Kiap patrols. Otherwise they had only stone tools and implements.

The women wore grass skirts while the men covered themselves with a loin cloth in the front, and over their backsides a hanging bunch of leaves known evocatively in Pidgin as 'arse grass'. They exist on a diet of root crops, which they grow in their forest 'gardens', augmented by whatever they can hunt down with their bows and arrows of sharpened bamboo. Their precious domestic pigs are only slaughtered on very special occasions and are otherwise treated with loving care. The women walk around clutching piglets as if they were their own children and will even suckle them if necessary.

Friendship having been further cemented with gifts of salt and tobacco, it was agreed that the patrol would spend the night in the village before clearing a site for the re-supply drop the next morning. The villagers watched spellbound as Dave Weaver prepared the powerful Plessey Clansman radio and called me up at Operation Drake's TAC HQ 350 miles away in Lae.

In the course of his message – the main purpose of which was to give the patrol's position so that the re-supply plane would know where to find them – he naturally mentioned that they had come across a tribe that had never seen a white man before. No sooner had he signed off 'Roger and out' than the news agency lines were buzzing with a typically sensational out-of-context story that caused us all a certain amount of embarrassment.

According to the headlines in the world's newspapers the following morning 'a lost tribe' of 'stone-age' savages had been discovered in 'the land that time forgot' where 'headhunting and cannibalism are still rife'! It turned the whole episode into something only marginally less miraculous than if Conan Doyle's 'Lost World' had been found complete with real live dinosaurs.

One couldn't blame the papers – it was, after all, a good story. But it wasn't quite that good. And the newly-independent Papuan government were far from pleased by the implication that their country was still struggling in the neolithic age. They were particularly upset by the suggestions of cannibalism and headhunting.

In all fairness, Roger and his patrol never actually tried to claim that there was anything that extraordinary about the fact that there are still people living in isolated parts of the jungle who have never seen white men. As white men rarely venture into such parts it would be surprising if there weren't.

It was, nevertheless, a fascinating encounter and it is a pity that the media made just a bit too much out of it. The 'local colour' added by the references to cannibalism was most unfortunate – but quite understandable. There have been recent cases of people eating human flesh in Papua New Guinea – but there is a subtle difference between cannibalism and what is known officially as 'interference with the human body'. This latter offence involves eating flesh from a body that is already dead, as against going out and killing somebody specifically to eat him, and this was once common practice in Papua New Guinea as it was in several parts of the world. It is now outlawed, but old customs die hard and there is evidence that this one has not been entirely stamped out. Again, it has to be got into perspective.

After being lifted out of Tigaro in the Chinook – the idea of a simple supply drop had been changed because film man Al Bibby wanted to land and get footage of the 'lost tribe' and it then seemed silly not to make use of the transport – the patrol were dropped back on the Strickland at the Bulago Junction to continue their recce right to the 'Gates of Hell'.

The nearer they got, the tougher the going and they began to realise why Clancy had been content to view this mysterious spot from a distance. By the time they eventually reached it they were all in a sorry state. Dave Weaver's face was puffed and swollen from insect bites and his lips were cracked and bleeding so that he looked as if he'd just taken a hiding from Muhammad Ali. Bill Neumeister was covered in cuts, sores and a blistering rash caused by brushing against a particular nettle-like leaf and Bob Woods' arms and legs were a mass of festering tropical ulcers resulting from open sores that will not heal in the humidity. Roger had been bitten by a spider on his hand and this had caused his whole arm to swell up, he had an abrasion down his right shin which refused to heal and, to cap it all, he had a bad dose of diarrhoea. He wrote in his diary towards the end of the trip: 'Frankly, all of us will be pleased to see the back of this miserably wet, tangled stretch of river. It is only cups of tea and a sense of humour which keeps us going. Our main problem is that we have to cut every inch of the way with machetes. There are no trails on the steep, muddy banks because no one has ever been here before. And, as I listen to the torrential rain on the waterproof sheet above my hammock, I can understand why!'

Despite this he started making plans for a return visit. The verdict at the end of the recce was that the proposed river running venture was perfectly feasible. It would be hard and there would be times when they would have to get out of the water and by-pass some of the worst rapids, but as long as everybody knew what they were doing and didn't get careless there was no real reason why they shouldn't get through.

Having assembled back at Lake Kopiago, the two four-man teams who would be crewing the Avon inflatables British Caledonia and Cathay Pacific, went into white water training on the Watut River. Roger, Bill Neumeister and Dave Weaver had now been joined by Jim Masters and Yogi Thami – a young professional river runner from Nepal. The party was completed by American Jerome Montague, a Papua New Guinea Defence Force warrant officer and film cameraman Warwick 'Waka' Atwell.

Both Roger and Jim had had some harrowing white water experience during the Blue Nile Expedition of 1968, when one man was drowned and all of us had some close shaves, so no one needed to tell them that a fast-flowing river can turn into a killer the instant you relax. They had learned the hard way that you have to treat rivers like the Strickland with the greatest respect and they made sure that everyone else on the team knew exactly what to expect. There were plenty of thrills and spills on the Watut and by the time the team were ready to pick up their boats and walk in to the gorge to start the descent in the last week of October, none of them had any illusions about a quick paddle down a picturesque mountain stream.

The first few hours were easy enough but then the white water stretches began to get gradually more and more severe and soon the adrenalin was flowing as fast and furiously as the water. Before long they were going into rapids that, on the recce, had been rated 5 on the scale of 1–7. There was always the knowledge, too, that however closely a rapid is surveyed in advance, you can never really tell what it is going to be like until you are in it.

The first major mishap came when Jim Masters' boat just slipped off the tongue of clear water that arrows in to the start of any rapid. It immediately turned sideways and then capsized, spilling the crew into the boiling waters. It all happened so fast that cameraman 'Waka' was still filming as he went under. For a few seconds he bravely held on to his camera, but then realised that he would have to sacrifice it in order to save himself and several hundred pounds worth of equipment plus what would have been a very exciting clip of film disappeared for ever.

Meanwhile Jim Masters had also gone under and had come up again underneath the upturned boat. He could hear Dave Weaver anxiously shouting out his name, momentarily convinced that Jim was lost, and memories flooded back of a particularly horrific experience on the Blue Nile. He had been hurled into an exceptionally violent stretch of water under a small waterfall and the forces created by the turbulent whirlpools and swirling eddies sucked him down and pinned him to the river bed. He said afterwards that it was like being trapped in a giant tumble drier. He lost control of his arms and legs as he was rolled over and over and bounced helplessly against the boulders and rocks on the bottom. As consciousness began to fade he somehow managed to locate the ring-pull on his life-jacket and this provided just enough buoyancy to pull him clear and throw him to the surface. But he knew he had been close to death.

This time it wasn't nearly so desperate. Surfacing under the boat was the best thing that could have happened to him – even if it did give his colleagues a scare – because it not only gave him something to hang on to but also provided an air-pocket.

After that there were no more accidents until the one mighty catastrophe that brought the expedition to a final, premature halt. And that was caused by the one tiny lapse in concentration, the one moment of carelessness that they had all known could be their undoing and had tried so hard to avoid.

It happened shortly after they had negotiated the Kagwasa Crossing. This in itself had been fairly nerve-wracking since they knew as they approached that if they once got sucked into the current and funnelled into the fissure it would be worse than going over Niagara in a barrel. They avoided that and it was the next day, as they were 'lining' the boats through an impossibly difficult stretch when disaster struck.

Where they went wrong was in pushing themselves just a bit too hard. They had made a very strict point up until then of always packing it in for the day by 3.00 pm, specifically to avoid the risk of getting overtired and making silly mistakes. But on this occasion they broke their own golden rule and carried on until after 6.00 pm. The reason was that they wanted to get that difficult section behind them so that they could look forward to some easier going the next morning.

It was pouring with rain as they rigged the line and sent the first boat down the 500 yard rapid. It went through like a dream. This helped further to lull them into a false sense of security and they hurried on with the second boat. The mistake they made was to leave just a bit too much slack in the line attaching the bow of the boat to the carabina attachment on the bank. As a result, the boat was able to turn sideways on, whereupon it got swamped by a wave and began to fill with water. The extra weight put an intolerable strain on the 5,000 lb line which eventually snapped with a twang like a breaking guitar string. The boat was swept away in an instant.

As the team dismally took stock of the situation round their spluttering camp fire it soom became obvious that they couldn't go on. They thought briefly about pressing on with all eight men in the one boat but, although they could possibly have got through that way, they risked getting into really serious trouble if that boat were accidentally lost as well. They reluctantly decided that the only thing to do was to sit tight and wait for the helicopter that was due to make a re-supply drop to them in three day's time.

Just to make things worse, their radio was on the blink so they couldn't call the chopper in any earlier. On top of that, a considerable proportion of their remaining supplies had disappeared down the Strickland in the boat and they were very short of food. And to cap it all, poor Jim Masters had lost everything except the shorts and shirt he stood up in – and they were soaked. In the end he was reduced to a rather macabre form of dress. The party was equipped with black plastic body bags – just in case anybody should get killed – and Jim cut arm holes in one of them and walked around looking for all the world like some ageing punk rocker. It's not often that this veteran explorer has found himself so much in the height of fashion. All he needed to complete the picture was a safety pin through his nose!

But nobody could see the funny side of it at the time. The whole team were totally depressed as they bedded down for the night on the convenient flat ledge that was to be their home for the next three days and lay there in the dark and the rain listening to the eerie crash of boulders avalanching down the crumbling cliffs on the other side of the gorge.

With certain members of the party, disappointment at the failure of the mission gradually turned into festering discontent during the following few weeks when, having been airlifted from their ledge, they linked up with a group of five YE's and four scientists at Nomad, as planned, and progressed down the lower reaches of the Strickland carrying out a number of scientific research projects on the way.

From a psychological point of view it is an interesting fact that the two men who seemed to have the greatest difficulty in coming to terms mentally with the set-back were in fact those who were physically the strongest and most resilient of the team by far. One of them brooded over it to such an extent that he detached himself from the rest of the party and always set up his hammock apart from the main camp. It is a syndrome with which all expedition leaders are familiar. I remember on the Darien Quest that it was the tough guy of the team who cracked up first. On Operation Drake as a whole we found time and again that when the going was really tough it was the girls who seemed better able to cope. The simple explanation, presumably, is that for people whose physical strength and single-mindedness have always enabled them to win through, the taste of defeat is that much harder to stomach.

For all the eight members of the white water team the month-long cruise down the Strickland, into the Fly and then on to Daru, where they linked up with the *Eye of the Wind* and sailed on to the Papuan capital Port Moresby, must have seemed a little bit of an anti-climax. But for the scientists and YE's there was plenty of interest. Five different scientific camps were set up at intervals along the way and at each of these between two and five days were spent studying the local flora and fauna.

Much of the research centred around the area's considerable crocodile population and Jerome Montague's midnight croc hunts could always be relied upon to enliven the proceedings. It is no job for the faint-hearted. Crocodiles are basically nocturnal creatures and, as Jerome explained, the only way to count them is to paddle down the river in the dark, shine powerful torches along the mudbanks until you can see the reds of their beady little eyes and then divide by two. All very well, until one or two of the ugly brutes aren't mesmerised by the lights, as they are supposed to be, but take fright and dive under – or, on one never-to-be-forgotten occasion, right through – the boat. Sharing a small inflatable with a panic-stricken eight-foot crocodile in pitch darkness is the kind of experience that soon sorts the men from the boys!

Almost as fearsome as the crocodiles were the crocodile hunters – particularly the Australians. Some of these were characters straight out of fiction who – after years of living in jungle camps – had 'gone troppo', as their more sophisticated fellow-Aussies put it. At the Lake Murray Crocodile Project Centre – where the crocs are farmed – the hunters would arrive each morning with their catch and the live crocs – rendered harmless by having their vicious jaws tied together –

would be sold for so much per foot. When you consider that the approved method of catching a croc is to go up to it, shine the trusty torch in its face and then slip a wire noose over its snout, it no longer seems surprising that most crocodile hunters are, by nature, somewhat eccentric.

Meanwhile, I was planning to hunt a very rare beast indeed – the fabled dragon of Papua New Guinea.

Chapter 8

Many years ago a young Papuan warrior staggered breathlessly in a state of shock home to his village after a lone hunting trip and blurted out an amazing story which his grandchildren still recall vividly and which they will repeat, word for word, to anyone who penetrates far enough up the Binatori river to reach their isolated but idyllic village of Giringarede.

It seems that grandad was feeling rather weary after hours on the trail and went to sit down and rest on a fallen tree trunk, only for the 'tree' to rear up under him and reveal itself as a dragon! It stood over ten feet tall on its hind legs and had wicked-looking jaws like a crocodile. Grandad didn't wait to see if it breathed fire at him – he ran for his life and never once looked back to see if the monster was pursuing him.

It all sounds a bit far-fetched, until you consider that stories about creatures fitting very much the same description have been coming out of Papua New Guinea since the end of the last century, and many of them have come from very reliable sources. During the war, Allied and Japanese patrols, operating deep in the most remote parts of the jungle, reported catching glimpses of what was most often described as a 'tree-climbing crocodile'.

An Australian called K. R. Slater – at one time an animal ecologist with Papua New Guinea's Department of Agriculture, Stock and Fisheries and later Senior Wild Life Officer of the Fisheries and Game Department of South Australia – was so intrigued by the legends that he started his own investigations in 1952 and wrote an article in which he stated that once he began looking into the matter his original scepticism vanished. He came to the conclusion that what people had seen was a giant lizard of the type first officially identified in 1878 as Salvador's Monitor – a close relation of the famed Komodo Dragon of Indonesia. A number of very large specimens of Salvador's Monitor have been caught in Papua over the years. One, seven feet long, was trapped near the mouth of the Fly river in 1936 by a scientific expedition whilst a trader named John Senior – who runs a general store on the Kikori river – has a skin nailed to his wall which, in life, must have measured a good ten feet. On this evidence there seemed every reason to suppose that somewhere in the most inaccessible recesses of the jungle there might lurk one or two outsize freaks – the equivalent of those 'grandfather' pike and other big fish that anglers sometimes hook from deep, dark pools. As Slater wrote 'Stories of specimens of fifteen feet seem reasonable. It just remains for somebody to produce one.'

That was a challenge I found hard to resist and when the recce party I sent out came back with photographs of a seven-footer and reports of others even bigger, I made up my mind to go ahead and mount a search for what would undoubtedly be the longest lizard in the world.

I decided to base the expedition on Masingara – the home village of Somare Jogo, the liaison officer seconded to Operation Drake by the Papua New Guinea government, just up the coast from Daru, in the area between the Fly river and the Pahotori river which was generally considered the most likely haunt for any monster Monitors.

Things got off to a suitably dramatic start even before I and my twenty-strong patrol reached Masingara when we managed to get lost in some of the most treacherous waters in the world. The Coral Sea is a wonderfully romantic name for a bit of ocean that is full of deadly reefs, strong currents and shifting shallows. We never really gave this a second's thought as we set out from Port Moresby, bound for Masingara in the *M.V. Andewa*. The little Defence Force vessel and her crew had been put fully at our disposal by the government – another example of the incredibly generous help and co-operation given to Operation Drake throughout our stay in the island.

The three-day voyage began well enough. Everybody was sunbathing on the after-deck, while dolphins played in the bow wave and large, juicy baramundi obligingly swallowed the hooks which the crew trailed out astern and assured us of some excellent eating. It was around midnight on the second day that things went wrong. There was no sign of the lighthouse which was a vital landmark and, although everybody aboard strained their eyes through the darkness, it was to no avail. At this point the skipper casually confessed that he never bothered with charts because he knew the waters he normally sailed like the back of his hand, but that this part of the ocean was quite new to him and he really didn't know where we were! After haphazardly changing course a few times in the hope of hitting upon some identifiable landmark our position was even more confused. I did have a chart with me but it wasn't a lot of help when we couldn't pinpoint our location.

We were literally going round in circles when someone brightly suggested that as we were in such dangerous waters it might be a good idea to throw a lead line overboard just to be on the safe side and this revealed the startling fact that we were over a sandbank with less than two feet to spare! The skipper inched into slightly deeper water before anchoring for the night in the hope that we would be able to get our bearings in the morning when we would at least be able to see what we were doing.

Dawn broke a few anxious hours later to reveal that the water, which the day before had been deepest blue, was now a murky, muddy brown while large clumps of natural flotsam and jetsam – including whole palm trees – floated by in a strong current. This gave us the vital clue that we must be somewhere near the mouth of the mighty Fly river and enabled us to chart a course which eventually led us back to safety. Even so, there were some more tense moments before someone spotted the faint outline of an island on the horizon which made it possible for us to fix our position exactly.

No sooner had we anchored off Masingara and marched the half-mile inland to the well-ordered village of traditional stilt-supported bamboo huts than I was ushered towards a 100-

year-old woman who was the most senior citizen and who was said to know more than anyone about the dragons having seen several in her lifetime. The white-haired old lady confirmed many of the things we had already heard – that these creatures grew to over 15 feet in length, that they often stood on their hind legs and so gave the appearance of dragons and that they were extremely fierce.

This last point brought much nodding from the village hunters who made it quite obvious that they treated even the smaller six or seven-footers – which they said were quite common – with the greatest respect. This came as no surprise, since we had already been told of an incident in another village where a captured Monitor had smashed its way out of a stout cage and killed a large dog before escaping back into the forest. Now we learned that the lizard's method of hunting was to lie in wait in the trees before dropping onto its victims and tearing them to shreds with its powerful claws. Apart from that it possessed a very infectious bite – the result of feeding on carrion – which could bring death within a matter of hours. And everybody had heard of cases where men had been attacked and killed by lizards.

During the next few days we split into four patrols and combed the surrounding jungle, but although everyone we met understood immediately when we explained what we were looking for and claimed that they themselves had seen such creatures the nearest any of us got to a sighting was when the village dogs that were with my patrol put up something that crashed off heavily through the undergrowth without showing itself. The patrol that ventured up to Giringarede and heard about the dragon disguised as a tree, got the local hunters to take them out to the spot where the incident occurred but found nothing. There was a flurry of excitement when they returned to the village to be told that the dogs had killed a big lizard while they were away – but it turned out to be a three-foot iguana.

At this point I decided to introduce an element of bribery into the proceedings, a tried and tested method of raising enthusiasm for any such hunt. I have never forgotten the lesson of Ethiopia in 1964 and the search for Osgood's Swamp Rat. We had had absolutely no joy at all until I offered the local hunters a reward and then suddenly specimens of every living creature in the country started arriving in camp strung on poles. As it happened, Osgood's Swamp Rat was unfortunately not among them. But that wasn't the hunters' fault. As we discovered when we returned empty-handed to England, the British Museum had sent us on what was almost literally a wild goose chase. It was hardly surprising that we couldn't find the elusive rodent since it lived not in Ethiopia but in the Congo.

Anyway, the point was that the promise of a tenner to anyone who brought in the wanted creature dead or alive had certainly brought results – even if they were the wrong ones! And since there could be very little doubt about what we were after on this occasion there seemed a good chance that we might get lucky. Sure enough, the mere mention of money was followed by a mass exodus of every able-bodied man in the region into the jungle armed with everything from bows and arrows to an antique blunderbuss.

It was as we steamed back down the Pahotori in the *Andewa*, after a trip up river that had produced many more confirmations that what we were looking for did indeed exist but no actual sightings, that the news came over the radio that, deep in the forest, a hunter had managed to shoot a big Monitor and was on his way back to Masingara with it. We returned to the village at full speed and by the time we got there a large crowd had gathered in a circle around the lizard which was lying at their feet roped to a bamboo pole.

It was no dragon – and yet it was still a pretty fearsome-looking specimen at just over six feet. And when our zoologist, Ian Redmond, had performed his post mortem he announced that it was only a baby – so one was still left speculating about what an overgrown adult might look like. Meanwhile, a small patrol that had been keeping vigil beside a remote water hole, which we had been told the Monitors frequented, came back with reports that they had seen several quite sizeable ones coming down to drink at night but had been unable to get near enough to photograph them in the dark.

By the time we had to leave we had still not seen anything that could be thought of as a dragon – but we were even more convinced that such giants did exist. Ian Redmond was so excited by the possibility that he stayed on after the rest of us had gone and made several sightings of impressive specimens, including one estimated at twelve feet. I am still confident that had we had more time to penetrate deeper into the jungle where no one ever goes, we could well have met up with the kind of monster that gave the man from Giringarede such a fright. A further, better-prepared attempt to flush out such a wonder is high on my list of future projects. Meanwhile one of our Young Explorers – Dorian Huber from Switzerland – is already planning to mount an expedition of his own to track down a larger version of the Monitors he photographed for us during the recce.

All sorts of mysteries and secrets lie hidden in the silent, out-of-the-way corners of the Papua New Guinea jungle and some of them have a rather grim and macabre aspect as one of our patrols discovered. They had gone up into the Finisterre Mountains area to search for the wrecks of some of the 325 Allied aircraft that disappeared without trace during the fierce fighting of the Second World War. Even large bombers vanished for ever as the forest canopy closed over them and the dense undergrowth rapidly cocooned them in natural camouflage. Teams led by Mike Gambier, James Horlick and American Robbie Roethenmind succeeded in locating the remains of several of these mechanical monsters including a P47 fighter, a rare B-16 bomber and a DC-3 troop carrier which assumed a particularly eerie and haunting atmosphere for Robbie and his YE's when they found human bones inside the fuselage.

The coastal waters of Papua New Guinea also harbour a multitude of war debris as the diving team led by Captain Tony Molony soon discovered. They monitored four areas – Salamaua, Madang, Finschhafen and Rabaul – where they located and charted over forty wartime ship and aircraft wrecks.

The full horror of jungle warfare in an environment so hostile that more men were lost through disease than through enemy action was borne out most forcibly on those patrols that investigated the old wartime trails across the island – in particular the Bulldog and Kokoda Trails. The teams led by Lieutenant Desmond Monteith and Lieutenant-Colonel Robin Jordan got a very good idea of the problems and hardships that the Allied and Japanese troops must have faced as they sweated through one of the most unpleasant – and vital – campaigns of the entire war. The Japanese came within five miles of over-running the entire island – which would have given them a bridgehead to Australia – and were only thwarted in the end by the dogged resistance of the Allies and the impossibility of maintaining adequate lines of communication through the jungle.

The scientific projects in Papua New Guinea were focused around the aerial walkways at Buso Camp, some sixty miles south of Lae on the east coast. The two walkways – the second of which was left in place after Operation Drake left so that the Bulolo Foresty College could continue the studies – were constructed under the supervision of Royal Engineers Sergeant Louis Gallagher and, as in Panama, proved to be a tremendous boon for the scientists who revelled in the unaccustomed access they got to the forest canopy.

A team from the Queensland Institute of Medical Research carried out projects on parasitology and insect carriers of human disease, while the botany group from the University of Queensland studied the gaseous uptake of forest plants. Other scientists studied the distribution of birds, moths and butterflies and made collections of a wide-ranging variety of specimens for the Zoological Museum of Papua New Guinea and the British Museum.

Medical teams carried out a series of surveys in three coastal villages near Buso and also at various places in the West Sepik Province wherc they were able to undertake research specially requested by the local Provincial Health Department.

The Buso Camp – smoothly and efficiently run by Captain Anthony Evans of the Coldstream Guards – was one of the happiest and most productive of any of Operation Drake's worldwide projects with scientists and YE's working together harmoniously on a programme which yielded very definite practical benefits for Papua New Guinea as well as satisfying the need for original research and a bit of adventure.

Meanwhile, the *Eye of the Wind* – when she wasn't in dock under-going yet more expensive repairs – was used in a project based on the islands off the east coast that involved a survey of traditional medicinal plants by a team from the University of Papua New Guinea. A botanist from the National Herbarium collected eighty-five such plants including those used in the treatment of malaria and as contraceptives.

When Operation Drake finally packed its bags in Papua and moved on to Sulawesi it was with the gratifying feeling that we had not only learned a little more about 'the Great Unknown' but had also helped her to find out a bit more about herself.

Chapter 9

Stark, spine-tingling terror froze Dave Smith into corpse-like stillness in his sleeping bag as the awful realisation dawned as to exactly what it was that had woken him in the middle of the night. He had thought at first that it was indigestion that was sitting so heavy on his stomach – that evening's compo rations had been rather stodgy, after all. But as his hand moved across his belly it touched something smooth and leathery and instantly he knew with sickening certainty that his ultimate jungle nightmare had come true. There was a snake in bed with him. And, judging by the thickness of the coils he had briefly fingered, it was a big one.

For a few seconds he lay there in the pitch darkness of the palm-thatched village hut hardly daring to breathe. He couldn't hope to slide out of the enveloping sleeping bag without disturbing the monster. There was only one thing to do. Galvanising himself into action with a yell of 'Christ!' he somehow managed to leap from the bag, grab the snake and hurl it as far away from him as possible all in one swift movement. The sudden commotion had the other members of the patrol who were sharing the hut with him sitting bolt upright, wanting to know what the hell was going on. When Dave, still shaking like a leaf, explained what had happened, they searched the hut by torchlight – but there was no sign of the snake. Dave sat up wide awake for the rest of the night with a torch in one hand and his machete in the other wondering why he had ever allowed Wandy Swales, his neighbour back in Alnmouth, Northumberland, to talk him into taking temporary leave from his job as a hotelier to go and cook for a bunch of mad explorers deep in the Indonesian rain forest. He must have been mad! He never did find out exactly what kind of snake it was that found the warmth of his sleeping bag so attractive – but he is convinced that it was something deadly. And he still shudders when he thinks what might have happened had he turned over in his sleep!

Snakes are to Sulawesi what coffee is to Brazil – there's an awful lot of them. What's more, they tend to be bigger, uglier and nastier than those you find anywhere else in the world. Shortly before Operation Drake arrived on the island there was an great excitement when a 20-foot python was killed with the body of a man still intact inside it – and there were pictures to prove it. Less well-authenticated was the report from neighbouring South Jawa of a 22-foot snake that attacked a man in a bulldozer and fought a lengthy battle with the machine before being dislodged from the cab and cut up under the tracks. An estimated 18 foot python was seen swimming in the river near one Operation Drake camp site and an even larger one was brought in, dead, having been cleanly shot through the eye with a poisoned dart from a blowpipe. So, one way and another, it is not the kind of place one would recommend to anybody with a phobia about creepy-crawlies.

As far as phase leader Derek Jackson was concerned, snakes were the least of his worries – even though he does happen to have a horror of the creatures. Setting up the Sulawesi operation involved greater problems of logistics and organisation than perhaps any other phase of the whole two-year expedition. It is probably just as well that the ultimate turn-on for any explorer worth his salt is when people purse their lips, shake their heads and murmur with an air of finality, 'It can't be done'. Because it was with that gloomy forecast ringing in their ears that Derek and his second-in-command Wandy Swales faced the challenge of making Sulawesi work and triumphed to the extent that it ended up as one of the most popular and successful of all the phases.

Just getting there in the first place was difficult enough. When it comes to internal communications, Indonesia generally, and Sulawesi in particular, makes even Papua New

Guinea seem sophisticated by comparison. Roads are virtually non-existent and there is no network of light aircraft routes.

When Wandy was despatched by Derek on an advance recce aimed primarily at finding ways and means of reaching the remote village of Morowali – the spot which had been picked as a likely base camp site – he soon found himself involved in an epic journey. Of the month that he was away, all but a few days were spent either travelling or arranging to travel. It started smoothly enough with an RAF VC 10 flight to Hong Kong and then an onward connection with Cathay Pacific to Djakarta. He was sidetracked there when he got involved in the local Highland Games – an amazing thrash laid on by the local St Andrew's Society and proudly claimed to be the biggest event of its kind outside Scotland where he made a lot of vital contacts. Then it was on to Ujung Pandang – capital of Sulawesi – and from there to Palu with the Indonesian international airline Garuda before catching a Twin Otter light aircraft down to Poso. There he managed to hitch a lift on a jeep as far as Tentena from where an airborne Pentecostal missionary named Paul Huling – known locally to one and all as Pilot Paul – agreed to fly him down to Betelemi. He then set out to walk the 20 miles to Kolonadale but was eventually 'taxied' on the back of an ancient motorbike that took three hours to complete the journey along a track that the rains had turned into a quagmire. Finally he hired a dug-out canoe with an outboard engine that chugged him the fourteen miles across the bay to Morowali – at a cost of sixty pounds return! When he at last got there it was only to find that the village was several miles from where it was supposed to be according to the maps – a relatively incidental matter explained by the fact that the villagers liked to move around a bit and always took the village with them.

Wandy's wanderings left him in no doubt about the complications that were going to be involved in moving personnel and over seventy tons of stores and equipment into the area. Clearly, the itinerary he had followed was totally impractical – if they had to come that way the chances were that half the YE's would never actually arrive at all. After much frantic re-appraisal and seeking after local knowledge, it emerged that a much more straightforward alternative was to fly from Djakarta to Kendari as directly as possible before going down to the quayside and, for about five pounds per person, chartering a fishing boat for the three-day voyage up the coast to Kolonadale.

Although this was obviously much simpler it still provided a testing introduction to the delights of Operation Drake for those YE's who didn't come in the easy way aboard the *Eye of the Wind*. That part of the journey from Djakarta to Kendari tended to be plagued with delays and hassles as one waited for the opportunity to aquire standby seats on the few flights going, while the conditions on the boats up to Kolonadale ranged from uncomfortable to downright hair-raising.

There were never any proper toilet facilities on board and the only food available was that which the crew lived on – fish heads and rice soaked in hot chilli sauce. The boats would often be overloaded to the point where they were wallowing along with no more than a few inches of freeboard so that if the sea got up it could be very dangerous. This was brought home forcibly and tragically when one of the leakier old tubs turned over and sank with the loss of several lives during her very next trip after coming up with a load of Operation Drake people.

Inaccessibility was by no means the only factor that made life extra difficult for Derek and his team as they struggled to get the Sulawesi phase off the ground. There was also the frustration of having to deal with a ponderous and multi-tiered bureaucracy that demanded permits in triplicate for even the most minor formalities. On top of that, the political situation meant that one had to tread very warily.

It is hardly surprising in the circumstances that poor old Wandy Swales keeled over with a minor heart attack just as the final preparations were reaching a climax. Wandy, a forty-four-year-old North Shields businessman and ex-paratrooper who had been specially recruited by Derek as his second-in-command because of his toughness and his organisational expertise, had gone for two weeks immediately prior to the attack without ever having more than a couple of hours sleep a night. That was the kind of pressure he was working under. Happily, his collapse turned out to be not too serious and he was back on his feet within a matter of days.

Despite the many problems everything was all right on the night and the scientists and YE's arrived to find a relatively luxurious establishment awaiting them at the base camp, built in a jungle clearing near the Ranu river. Camp Ranu consisted of four large sleeping huts, each capable of accommodating up to sixty people, as well as a science hut, survey hut, medical hut, signals hut and cookhouse – all of them solidly constructed out of wood with roofs of thatched leaves. A very cosy set-up – and why not? There was a lot of serious scientific work to be done in Sulawesi and the environment guaranteed that life would be sufficiently uncomfortable without introducing unnecessary hardships.

The original reason for going to the Spice Islands was supplied by Sir Francis in January 1580. Having made a trade agreement with the Sultan of Ternate and somehow having found room to cram six tons of cloves into the already bulging hold of the *Golden Hind*, he set sail on a westerly course, bound at last for home on what he no doubt hoped would be an uneventful last leg of his long voyage. It was not to be.

At precisely 8.00 pm on the evening of 9 January, with the little ship creaming along under full sail before a fresh trade wind, there was suddenly a sickening, timber-jarring crunch and everybody aboard was thrown violently forward as the vessel's progress was brought to a grating halt. It was instantly obvious to one and all that she had fallen foul of that lurking menace that every seaman dreads – a submerged and uncharted coral reef.

The initial reaction was that it was all over – that it would only be a matter of minutes before the ship broke up and went to the bottom along with her crew and her precious cargo. However, a hurried inspection showed that she had miraculously survived the impact against the jagged rocks without being holed. Even so, she was stuck fast and there didn't seem a lot of hope. Drake threw eight cannon and half the spices overboard in a desperate attempt to lighten the stricken ship but she remained hard aground. Characteristically, Sir Francis refused to jettison any more of the spoils that were weighing her down. He preferred to take the risk of never going home at all rather than take the safe course and get back empty handed.

Fortune favours the brave and just as the *Golden Hind* seemed about to keel over a sharp gust of wind from an unexpected quarter was just enough to dislodge her and refloat her in deep water. The only person who didn't have cause to feel relieved was the ship's chaplain. He had been prophesying disaster and blaming it on Drake's past sins, so the miracle of salvation rather took the wind out of his sails. Drake excommunicated him in front of the assembled crew – claiming that as the Queen's representative on board he had the power to do so. Once his humour had been restored, he relented and re-instated the unfortunate cleric.

The prospect of locating and raising the *Golden Hind*'s discarded cannon was an exciting one – although by itself it did not justify mounting a whole phase in the area, but we were soon offered a grander purpose which fitted in perfectly with the aims and ideals of the expedition.

Firstly, John Blower – an old friend of the Scientific Exploration Society who lives in Indonesia where he is attached to the Directorate of Nature Conservation – pointed out, when I approached him for advice, that the government had recently decided to devote five per cent of Indonesia's total land area to Nature Reserves as part of an impressive conservation programme. This also included the development of several National Parks, one of which – in Morowali – was ready to go ahead as soon as the area had been surveyed and a comprehensive management plan produced. Any assistance that Operation Drake could offer would undoubtedly be most welcome.

This already seemed like a very good idea before Scientific Co-ordinator Andrew Mitchell clinched it by drawing attention to the fact that Sulawesi lay right on the Wallace Line – the naturalistic no-man's-land dividing South East Asia from Australia, where many unique and independent species of plant and animal life have evolved – and was therefore of paramount interest to specialists in every branch of natural science.

It seemed, then, that here was an ideal opportunity for Operation Drake to mix research, exploration and adventure in such a way as to end up with a positive contribution of great practical benefit to the host nation.

After consultations with John Blower and with the Indonesian authorities I asked Derek Jackson to lead the Sulawesi phase and to come up with a detailed programme of activities. Basically, the results of the scientific research and the socio-economic and medical surveys carried out with the help of the YE's would be used in the preparation of a Management Plan to be written by Dr Andrew Laurie, who was associated with David Attenborough's best-selling *Life on Earth* masterpiece. In addition, observations made by exhaustive YE patrols would be used by Royal Engineers Sergeant Phil Maye in the production of the first-ever map up to ordnance survey standard of the 2,000 square mile park area.

There can be no doubt that the overall sense of purpose surrounding the Sulawesi phase greatly enhanced the enjoyment of everyone taking part. As in Panama and Papua New Guinea the scientific projects were centred around the aerial walkway, erected on this occasion by Sapper corporals John Rimmer and Mike Prior. And once again, the extra dimension it added to forest studies was much appreciated by the botanists, biologists and ornithologists.

Leeds University entymologist Stephen Sutton – who has since raved about Operation Drake in Sulawesi as the best scientific expedition he has ever been involved in – was able to set up the longest-running zonal distribution experiment ever undertaken. It lasted through one entire lunar month and this is particularly significant since the moon is a key factor in determining insect activities. Butterfly expert Anthony Bedford-Russell managed to collect nearly 300 different species whilst British Museum beetle collector Martin Brendell amassed a 'bag' of literally thousands of creepy-crawlies. Batman Ben Gaskell recorded 450 specimens of which he collected and preserved around 300.

Bird and reptile expert Bill Timmis, from the Harewood Bird Garden in Yorkshire, spent endless happy hours observing such quaintly-named specimens as the Dark-Chinned Fruit Pigeon and the Pink-Breasted Cuckoo Dove but also suffered one grave disappointment which, when he related the sad tale, brought smirks to the faces of all but the most sympathetic of fellow naturalists.

Whilst on his way to climb Mount Tambusisi he spotted the nesting place of an Imperial Pigeon and excitedly marked the spot so that he could return later to study the bird's breeding behaviour which had never previously been observed. But when he hurried back enthusiastically to notch up this prestigious ornithological 'first', he was shocked to find the nest deserted and an ominous scattering of feathers underneath the tree. He returned, crestfallen, to camp only to be greeted with the awful revelation that the Indonesian guide Harim had also ear-marked the nest for future attention and had then nipped back shortly before poor Bill and bagged the bird for his supper! To make matters worse, he insisted on describing in some detail exactly how he had plucked it and boiled it up in a pot and how very tasty it had turned out to be.

While the scientists busied themselves with everything from the sex life of the shorea tree – painstakingly investigated by Dr Peter Kevan from Colorado Springs University and Swansea University's Dr Andrew Lack – to Australian Chris Watts' probe into the ancestry and evolution of the Sulawesi rat, the YE's were plunged into a hectic and varied daily schedule aimed at ensuring that nobody would be able to claim afterwards that they had at any time been left hanging around with nothing to occupy them.

No sooner had they stepped ashore from the *Eye of the Wind* than they were split up into small groups and despatched to one or other of four local villages for ten days, armed with a questionaire to be answered by each member of the population as part of the socio-economic and medical survey. With no supervision from Operation Drake directing staff, the youngsters were given an early taste of the independence which was to be a much-welcomed feature of the Sulawesi phase. As a method of acclimatisation it was as successful as it was imaginative. The YE's reacted with great enthusiasm to the challenge of being thrown back on their own resources and the villagers were delighted with the novelty of having them. By the time they returned to base camp they had not only adjusted physically to living in the tropics but had also learned at first hand a great deal about their new environment and had even picked up a bit of the language.

Once acclimatised they were ready for a spell down at 'Eddie's Beach' – a special camp set up near the mouth of the Ranu river where they were taught the art of jungle survival by Eddie McGee. From Eddie – ex-paratrooper, Army Physical Training Corps instructor, and born survivor – they learned how to make rope out of grass, turn animal fat and hot water into soap, use ash from the fire as a sterile dressing and produce a gourmet meal out of nothing but ferns and bamboo. They also picked up such titbits of information as the fascinating fact that a brazil nut will burn like a candle for up to an hour, that a feather is unbeatable when it comes to cleaning out wounds and that a condom is probably the most versatile product known to man, serving every imaginable purpose from keeping one's matches dry to attracting the attention of passing aircraft – when blown up and sprayed with water the things become iridescent!

At the end of their week's training each YE was given the chance to put theory into practice by being dropped off alone on a remote desert island beach and left there for three days and three nights equipped with no more than a basic survival kit, a packet of spangles, a packet of biscuits and a litre bottle of water. Most of them admitted to being a bit twitchy on their first night, jumping at every little noise, but very few took the easy way out by sitting tight in one spot, ekeing out their meagre water and food supplies until the pick-up boat returned. The vast majority responded well to the thrill of real adventure and entered fully into the Robinson Crusoe spirit of things by building fish and animal traps, tapping fresh water from vines and searching for fruit. One young man spent hours longingly contemplating a coconut nestling way out of his reach at the top of a slender fifty foot palm before the answer came to him in a flash of inspiration – and he chopped the whole tree down with his machete! It had to be gently pointed out to him when he got back that although this showed admirable logic, it was not quite the thing to do in a conservation area!

The survival training and the 'solo' stint was far from being just a bit of fun. Derek Jackson, who has gained considerable experience of dealing with young people through his association with the British Schools Exploring Society – believed very strongly that the secret of success as far as the YE's were concerned was to leave them, wherever possible, to their own devices and to keep supervision to a minimum. Having put them through a rigorous training and acclimatisation programme he felt quite happy about following through with this policy even to the extent of sending patrols out deep into the jungle without any directing staff to chaperone them.

This added responsibility was gratefully accepted by the YE's who more than justified Derek's confidence in their ability to look after themselves. They coped very competently with everything they came up against – including a confrontation with the so-called Wild Wana. The Wild Wana are the native tribesmen of the interior whose mysterious ways and ability to fade away into the jungle and keep out of sight has earned them the name Kayu Merangka – which, literally translated, means 'leaves blown on the wind'. All over Indonesia they have a reputation as primitive, head-hunting savages who are likely to murder anyone who strays into their territory with poison darts from their blowpipes. Operation Drake had been warned to beware of them by the local authorities who insisted that they could be dangerous.

It was while on patrol North of the Solato river that a twenty-one-year-old cockney diamond sorter called Noel English awoke one morning to find two Wild Wana studying him silently and intently from the edge of the clearing where he was lying in his hammock. Once he had got over the initial shock, Noel put on his friendliest smile and advanced slowly towards the two men who were both carrying blowpipes. At first they retreated nervously and he realised that they were far more frightened of him than he was of them. Once he made it clear that he meant them no harm they inched warily back into the clearing and, when the ice had eventually been broken, they relaxed to the extent that they invited the whole patrol back to their village.

Here it soon became obvious that far from being murdering savages, the Wild Wana were gentle, shy people who had got their bad reputation simply because they were timid and shied away from contact with civilisation. And this was largely because they had had unhappy experiences in the past in their dealings with the outside world.

Another patrol of five YE's from Britain, America, Australia and Indonesia learned from a tribe of semi-settled Wana the secret of the poison with which they tip their blowpipe darts. All the Wana men carry blowpipes and although they normally only use them to kill birds and small animals it is said that the poison is so powerful that it can kill a man in less than a minute!

Normally, the source of this poison is kept a closely-guarded secret but as the YE patrol followed the course of the Sumara river in an attempt to discover, for the purposes of Phil Maye's map, whether it did eventually link up with the Ula river system in Central Sulawesi, they were befriended by the villagers of Langada who revealed the locations of two of the rare 'impoc' trees from which the poison was tapped. One was at a village called Malino – more than four days' march away – and the other was slightly nearer at a place called Bintana.

It took the patrol three days to reach Bintana. They were helped along part of the way by a guide who insisted on wearing his 'Sunday best' clothes – including a red tin construction helmet which looked rather incongruous but of which he was exceedingly proud. When they eventually got to the village they discovered the tree on the outskirts. Over 100 feet high and more than a yard in diameter, it was surrounded by bamboo scaffolding and the spongy brown bark was covered in scars and V-notches. When this bark was slashed with a machete it released a fountain of milky white sap, some of which the YE's collected very gingerly in a container for later analysis by the scientists. They returned to Camp Ranu highly excited at having discovered the secret of the Wana impoc tree.

While the YE's continued to explore the depths of the jungle with impunity, investigating every river and tributary and returning in triumph with ever more data for Phil Maye and Andrew Laurie, it was left to a patrol composed entirely of senior directing staff to get involved in the only real crisis of the phase.

The patrol – led by Sergeant John Cornish of the 14/20 Kings Hussars and the Junior Leaders Regiment of the Armoured Corps and including signaller Sergeant Bob Hooper, sapper Corporal Alan Bretherton and medical officer Dr Ian Gauntlett – set off for the remote region around the headwaters of the Solato and Morowali rivers in search of a 'lost world', a sheer-sided plateau with a lake on top of it and a waterfall fed by the lake which exited half way down the cliff before cascading 600 feet.

The party was accompanied by an Indonesian employee from Camp Ranu called Andreas and two Wana guides. Trouble started when Andreas fell out with John Cornish. At a time when everyone was tired and tempers were frayed, a violent argument developed over something really quite trivial and this ended with Andreas deserting the patrol during the night along with the two Wana guides.

The patrol were now left in a very tricky situation since they were in extremely treacherous terrain full of deep ravines, crumbling cliffs and hidden hazards where anyone who didn't know exactly where he was going could get into real trouble. They made it back in the end – but not without considerable difficulty and a fair amount of panic back at Camp Ranu when Andreas and the guides returned without them.

As Andreas was complaining hysterically that Cornish had tried to kill him the incident could also have caused ill feeling between Operation Drake and the locals, but this threat was happily avoided thanks to the presence of the official

Indonesian liaison officer, Colonel Johannes Prasanto.

Colonel Prasanto had been seconded to Operation Drake precisely to defuse these kinds of situations should they arise and he did the job superbly. He ended up more as a friend to the expedition than an official trouble-shooter and proved a tremendous asset in every way.

General Sir John Mogg was just one of a string of VIP's who visited Camp Ranu. Our man in Jakarta, Terence O'Brien, and his wife scored a big hit with everyone because of their enthusiasm – especially Mrs O'Brien, who insisted on being winched up to the walkway while other less adventurous dignitaries present made it quite clear that no way were they going to move one foot above ground level.

When the Indonesian Minister for the Environment, Mr Emil Salim, decided to make an official visit, Operation Drake got an object lesson in local bureaucratic protocol that brought the expedition to a complete standstill for thirty-six hours. The fact that Mr Salim was making an appearance automatically meant that all the regional heads of department had to be present and each of them had to bring his assistant and the assistants brought their servants and the servants brought their friends and suddenly food and accommodation had to be found for nearly 100 people. Chaos reigned and several of Camp Ranu's goats had to be slaughtered so that everybody could be fed with the result that a number of the lady YE's – who had grown greatly attached to the beasts – dissolved into tears.

The distress over the goats was as nothing compared with the great upset which developed around Daisy the cow. Daisy, a brown-eyed Jersey, was brought to Camp Ranu by some friendly locals and was swiftly adopted as a kind of camp mascot. However, not everybody was totally enamoured of Daisy – certainly not when she became so tame that she would wander nonchalantly through the dining area taking great mouthfuls of any morsel that took her fancy. You either loved Daisy or you hated her and a great controversy arose.

This reached a climax when those who objected to her suggested that she should be slaughtered and salted away in the *Eye of the Wind*'s deep freeze. Ship's engineer Bob Coupland made a big show of sharpening his butchery knives and there was a lot of lip-smacking anticipation of juicy steaks. It was all a bit of a joke really but it soon got beyond that as far as Daisy's supporters were concerned. The girls were in near hysterics but in the end it was the big, tough Australians who took the cow off and hid her in the jungle so that nobody could get at her. It all quietened down after that and Daisy was officially reprieved. In fact, she was still there, looking sleek and overfed, when Operation Drake finally moved on.

The Sulawesi phase ended with a major scientific seminar held in Jakarta under the auspices of the Indonesian Institute of Sciences and attended by many of the country's top scientists. It was a fitting conclusion to an expedition that had produced so many solid scientific achievements.

The only failure concerned the project which had started it all in the first place – the search for Drake's cannon. The magnetometer failed to pinpoint any 'target' on the Vesuvius Reef that was considered worth following up and the divers couldn't find anything either. But even here there was a consolation. Tricia Holdway confirmed that the reef was so rich in marine life that it should be considered an underwater conservation area and this suggestion has now been taken up by the local authorities.

Chapter 10

On Sunday, 20 July 1980 four Kenyan YE's taking part in the Masai Mara Game Reserve project in Kenya got into a Landrover and drove off to post some letters at the Keekorak Lodge hotel. On the way back to base camp the Landrover careered off the road on a corner, rolled over twice and ended up in a ditch. All four occupants were injured and one – Andrew Maara – died on the way to hospital.

We were still reeling from the shock of this disaster when a similarly horrible accident dealt us a second hammer blow of tragedy.

Richard Hopkins was one of six YE's who were returning from the ARK walkway site to base camp when their Landrover swerved to avoid a buffalo, careered off the road and turned over.

Robert, who was twenty and came from Yorkshire, died instantly.

When the terrible news came through we were all stunned. At the back of my mind there had always been the awareness that we must be prepared for at least one serious injury or illness and maybe even a fatality some time during Operation Drake. With so many people involved in adventures that were often demanding, occasionally dangerous and nearly always conducted in a strange and hostile environment it would have been extraordinary if we had excaped totally unscathed. But to lose someone in such a silly, unnecessary, commonplace way seemed somehow tragically ironic.

We had already had a scare which had demonstrated how the greatest disasters can so often be the least predictable. As the *Eye of the Wind* sailed up from Sulawesi to Kenya we got an SOS from the ship's doctor alerting us to the fact that American YE Cindy James had developed an abscess on the retina of her eye and was in urgent need of specialist hospital treatment. As soon as the ship arrived in Mombasa she was transferred to hospital in Nairobi. It emerged that there was a real risk that if the abscess burst Cindy's sight would be seriously affected and there were some very tense moments before it eventually subsided without causing any major damage. It had been a close thing – and all, it seemed, because of a dirty contact lens.

The full story of the Kenya phase will be told in a later book on Operation Drake which will cover the whole of the two-year expedition in much greater detail. It was actually substituted in place of the Sudan phase which unfortunately had to be cancelled at the very last minute owing to various political complications. Despite the need for some fairly frantic re-organisation it turned out to be anything but an anti-climax.

Projects were centred on four different areas. In the North a team of scientists and YE's carried out a survey of Lake Turkana while others either helped with the marking out of the boundaries of the Sibiloi National Park or went on a camel trek in the northern deserts.

In the central region around Aberdare work was carried out constructing the special Ark Walkway from which tourists can observe wild life whilst another party investigated the amazing Susua Crater which, despite being only some forty miles from Nairobi, is one of the few remaining corners of Africa which is largely unexplored. Now the Kenyan Wildlife Conservation and Management Department hope to turn it into one of their greatest tourist attractions – and Operation Drake was delighted to have the opportunity to undertake a recce aimed at investigating the feasibility of such a plan.

In the South, the Masai Mara Game Reserve project involved a detailed census of the wildlife as well as the building of cattle dips, while on the coast YE's combined with scientists and divers on archaeological and marine studies near Lamu and at Malindi.

The Kenya phase was finally wound up in early October and meanwhile the *Eye of the Wind* had already set out on the final leg of the expedition back to Plymouth.

As the great adventure drew to a close there was time to look back over the whole vast venture – and forward to the future. There was the tremendous satisfaction of knowing that, against all the odds imposed by limited finance and mind-bogglingly complicated problems of logistics and administration, Operation Drake had achieved most of the goals we set for ourselves. Our successes were not only tangible ones of discoveries made, research carried out, trails blazed and assistance rendered. Most important of all was the awakening of a spirit of enterprise, endeavour and initiative among more than 300 Young Explorers who were given a unique opportunity to explore the limits of their own capabilities and to discover the full range of their own potential.

The reactions we have had from all those who took part has encouraged us to consider ways and means whereby the ideals, principles and administrative network of Operation Drake can be kept alive. Our hope is to establish a Drake Fellowship as a kind of 'after-care' organisation which will not only maintain contact between all those involved in the expedition but which will also serve as a permanent body through which follow-up projects could be launched.

The first steps towards such a scheme have already been taken in Papua New Guinea and the Operation Drake committees in Australia, New Zealand, Hong Kong and Panama are also keen to carry on the good work and to build on the foundations that have been so solidly laid in the last two years. In the UK it is hoped to set up a small full-time staff selected from the Operation Drake team through which the worldwide activities of the Fellowship would be co-ordinated.

Among the future projects that are even now under active consideration is a return to Caledonia Bay in 1981/2 to continue work on the excavation of Acla and the raising of the *Olive Branch*. The idea is to re-activate the Operation Drake base camp for a further two years and take out young people, particularly under-privileged youngsters, for periods of two months.

Still more ambitious is the plan for another world-wide venture based on an ocean-going support vessel. This time we are thinking of a motorised ship of around 1,500 tons with a scientific laboratory, decompression chamber, a small educational broadcasting studio, stocks of tools and stores to be used in community aid projects, a medical centre and audio-visual equipment for lectures to schools in the areas visited. It is intended that young people would join the expedition for set periods, as on Operation Drake, but with the difference that they would be using sailing vessels from their own countries around which various local projects would be centred.

At this stage the scheme is still on the drawing board. There is much work to be done and a great deal of money to be raised before it can be put into practice. But if everything goes according to our preliminary schedule the expedition would be launched sometime in 1984. What an apt year, with all its gloomy connotations, in which to introduce hundreds more youngsters to the brave new world of adventure and discovery so thrillingly pioneered by Operation Drake.

GENERAL NON-FICTION

	0352307781	FRED SCHRUERS **Blondie**	**£1.25***
	0352307099	PEGGY ANDERSON **Nurse**	**£1.50***
	0352307498	DENYS VAL BAKER **The Spirit of Cornwall**	**£1.50**
	0352301392	LINDA BLANDFORD **Oil Sheiks**	**£1.50**
	0352307501	LT. COL. JOHN BLASHFORD-SNELL WITH MICHAEL CABLE **In The Wake of Drake**	**£1.25**
	0352396121	ANTHONY CAVE BROWN **Bodyguard of Lies** **(Large Format)**	**£2.50***
	0352307005	ROBIN COLLYNS **Prehistoric Germ Warfare**	**£1.25**
	0352306432	RODNEY DALE & IAN WILLIAMSON **Myth of the Micro**	**£1.50**
	035230345X	RODNEY DALE AND JOAN GRAY **Edwardian Inventions**	**£2.95**
Δ	0352301368	JOHN DEAN **Blind Ambition**	**£1.50***
	0352304413	MARGOT FONTEYN **A Dancer's World (illus)**	**£1.95**
	0426178653	RICHARD GARDNER **The Tarot Speaks**	**50p**
	0352303247	H. R. HALDEMAN **The Ends of Power**	**£1.25***

* Not for sale in Canada. • Reissues
Δ Film & T.V. tie-ins

GENERAL NON-FICTION

ISBN	Author / Title	Price
0426163443	XAVIERA HOLLANDER **Letters to the Happy Hooker**	80p*
0352307137	**The Happy Hooker**	£1.25*
0426166787	**Xaviera on the Best Part of a Man**	95p*
0352306688	**Xaviera Goes Wild**	95p*
042617996X	**Xaviera Hollander Meets Marylyn Chambers**	80p*
0352303891	ANNE FLETCHER **The Happy Hooker Goes to Washington (F)**	90p*
042700442X	WILLIAM KEMBLE **How to Pass Exams**	95p
035230300X	JAMES MARGACH **The Abuse of Power**	95p
0427004497	PATRICK MOORE AND IAIN NICHOLSON **Black Holes in Space**	75p
0352304790	JUDITH MIDGLEY-CARVER AND AMANDA DUCKETT **Career Choices**	70p
0427004284	JAY ROBERT NASH **Darkest Hours (Large Format)**	£5.00
035230653X	RACHEL NELSON **Success Without Tears**	£1.25
0352306165	MICHAEL NICHOLSON **The Yorkshire Ripper**	95p
0352302151	MOLLY PARKIN **Good Golly Ms Molly** (see also under General Fiction)	£1.25
0352303476	JEAN PLAIDY **A Triptych of Poisoners**	80p

* Not for sale in Canada. • Reissues
Δ Film & T.V. tie-ins.

GENERAL NON-FICTION

Δ	0352302410	ESTHER RANTZEN **That's Life (Large Format)**	£1.95
	0491024681	FREDDIE REED **The Queen Mother and Her Family**	£1.95
	0352398639	DONALD RUMBELOW **The Complete Jack the Ripper**	£1.25
	0352306394	ALEC X. SNOBEL **The Devil in Angela**	£1.25
	0352307684	PETER STONE **The Boomtown Rats (Large Format Illustrated)**	£2.95
	0426105249	NOEL STREATFEILD **The Noel Streatfeild Summer Holiday Book (illus)**	50p
Δ	0352307471	SIR FREDERICK TREVES **The Elephant Man**	95p
	0352306832	JOHANNES VON BUTTLAR **The UFO Phenomenon**	£1.25

MARLBOROUGH ISLAND GUIDES

0427004527	**Canaries**	95p
0427004535	**Channel Islands**	95p
0427004519	**Corfu**	95p
0427004543	**Isle of Wight**	95p
0427004551	**Majorca**	95p
0427004578	**Minorca and Ibiza**	95p

* Not for sale in Canada. • Reissues
Δ Film & T.V. tie-ins.

GENERAL FICTION

		CYRIL ABRAHAM	
Δ	042616184X	**The Onedin Line: The High Seas**	**80p**
Δ	0426172671	**The Onedin Line: The Trade Winds**	**80p**
Δ	0352304006	**The Onedin Line: The White Ships**	**95p**
Δ	042697114X	**The Shipmasters**	**80p**
		BRUCE STEWART	
Δ	0352305738	**The Onedin Line: The Turning Tide**	**£1.25**
		TESSA BARCLAY	
	0352304251	**A Sower Went Forth**	**£1.50**
		JUDY BLUME	
	0352302712	**Forever**	**75p***
		ANDRÉ BRINK	
	035230703X	**A Dry White Season**	**£1.95**
	0352305916	**Rumours of Rain**	**£1.95**
	0352306904	**An Instant in the Wind**	**£1.95**
		MICHAEL J. BIRD	
Δ	0352302747	**The Aphrodite Inheritance**	**85p**
		ADRIAN BROOKS	
	0352307773	**The Glass Arcade**	**£1.50***
		MAGDA CHEVAK	
	0352303514	**Splendour in the Dust**	**£1.50**
		JACKIE COLLINS	
Δ	0352395621	**The Stud**	**£1.25**
	0352300701	**Lovehead**	**£1.25**
	0352398663	**The World is Full of Divorced Women**	**£1.25**
Δ	0352398752	**The World is Full of Married Men**	**75p**

* Not for sale in Canada. • Reissues.
Δ Film & T.V. tie-ins.

GENERAL FICTION

	ISBN	Author / Title	Price
		CATHERINE COOKSON	
	0426163796	**The Garment**	95p
	0426163524	**Hannah Massey**	95p
	0426163605	**Slinky Jane**	95p
		HENRY DENKER	
	0352396067	**The Physicians**	95p*
	0352300523	**A Place for the Mighty**	75p*
	0352303522	**The Experiment**	£1.50*
		MEL ELLIS	
Δ	0352306386	**Wild Horse Hank**	95p*
		NORMAN GARBO	
	0352305665	**Cabal**	£1.50*
	0352304995	**The Artist**	£1.50*
		ROBERT GROSSBACH	
Δ	0352307528	**Chapter Two**	£1.25*
		ELIZABETH FORSYTHE HAILEY	
	0352304359	**A Woman of Independent Means**	£1.25*
		SUSAN HILL	
Δ	0352307242	**Breaking Glass**	95p
		BURT HIRSCHFELD	
	0352398582	**'Father Pig'**	95p*
	0352395176	**Secrets**	95p*
	0352398604	**Behold Zion**	95p*
		JOYCE JOHNSON	
	0352306718	**Bad Connections**	£1.25*
		SARAH KERNOCHAN	
	0352304200	**Dry Hustle**	£1.25*
		ELLIE LING	
	0352306416	**This Year's Girl**	£1.25
	0352304154	**The First Splash**	75p
		PAT McGRATH	
	0352304383	**People in the Crowd**	95p

* Not for sale in Canada. • Reissues.
Δ Film & T.V. tie-ins.

GENERAL FICTION

		LEE MACKENZIE	
Δ	0352396903	**Emmerdale Farm (No. 1) The Legacy**	**70p**
Δ	0352396296	**Emmerdale Farm (No. 2) Prodigal's Progress**	**80p**
Δ	0352395974	**Emmerdale Farm (No. 3) All That A Man Has . . .**	**75p**
Δ	0352301414	**Emmerdale Farm (No. 4) Lovers' Meeting**	**70p**
Δ	0352301422	**Emmerdale Farm (No. 5) A Sad and Happy Summer**	**70p**
Δ	0352302437	**Emmerdale Farm (No. 6) A Sense of Responsibility**	**75p**
Δ	0352303034	**Emmerdale Farm (No. 7) Nothing Stays the Same**	**75p**
Δ	0352303344	**Emmerdale Farm (No. 8) The Couple at Demdyke Row**	**75p**
Δ	0352304103	**Emmerdale Farm (No. 9) Whispers of Scandal**	**75p**
Δ	0352304510	**Emmerdale Farm (No. 10) Shadows From the Past**	**75p**
Δ	035230569X	**Emmerdale Farm (No. 11) Lucky for Some**	**80p**
Δ	0352304340	**Early Days at Emmerdale Farm**	**75p**

* Not for sale in Canada. • Reissues.
Δ Film & T.V. tie-ins

GENERAL FICTION

		JOAN MORGAN	
Δ	0352304367	**Mary Blandy**	£1.50
		N. RICHARD NASH	
	0352301562	**East Wind, Rain**	95p*
	0352395060	**Cry Macho**	£1.25*
	0352303778	**The Last Magic**	£1.75*
		ANAÏS NIN	
	0352302720	**Delta of Venus**	£1.25*
	0352306157	**Little Birds**	£1.25*
		ALAN PARKER	
	0352303271	**Puddles in the Lane**	70p
		GRAHAM PARKER	
	0427004470	**The Great Trouser Mystery (Large Format illus)**	£3.95
		MOLLY PARKIN	
	0352304533	**Fast and Loose**	95p
	0352300809	**Love All**	95p
	0352397179	**Up Tight**	95p
	0352302631	**Switchback**	95p
	0352302151	**Good Molly Ms Molly (illus) NF**	£1.25
	035230331X	**Purple Passages (illus poetry)**	£1.50
		JACK RONDER	
Δ	0352303581	**The Lost Tribe**	£1.50
		FIONA RICHMOND	
	0352398779	**Fiona**	75p
	0352303808	**The Story Of I**	75p
	0352305215	**On The Road**	75p
	0352305568	**The Good, The Bad and The Beautiful**	95p
	035230748X	**Galactic Girl**	95p

* Not for sale in Canada. • Reissues.
Δ Film & T.V. tie-ins.

GENERAL FICTION

	MARIA ISABEL RODRIGUEZ	
0352305622	**Maestro of Alhora**	£1.25
0352303913	**The Olive Groves of Alhora**	75p
	JUDITH ROSSNER	
0352396946	**To The Precipice**	85p*
0352302089	**Nine Months in the Life of an Old Maid**	75p*
0352301465	**Any Minute I Can Split**	95p*
	LAWRENCE SANDERS	
0352302135	**The Pleasures of Helen**	95p*
	JEREMY SCOTT	
0352307048	**Hunted**	£1.50
	ALAN SILLITOE	
035230698X	**The Storyteller**	£1.50
0352300949	**Men, Women and Children**	75p
0352300981	**Saturday Night And Sunday Morning**	£1.25
0352395141	**The Widower's Son**	£1.50
0352397144	**The Flame of Life**	£1.50
0352398809	**The Ragman's Daughter**	£1.25
035230202X	**A Start in Life**	£1.75
0352301821	**Raw Material**	95p
0352302518	**Key to the Door**	£1.50
0352303263	**The Death of William Posters**	£1.50
0352303379	**A Tree on Fire**	£1.35
0352305185	**Guzman go Home**	£1.25

* Not for sale in Canada. • Reissues.
Δ Film & T.V. tie-ins.